Communications in Asteroseismology

Volume 163
2011

Communications in Asteroseismology
Editor-in-Chief: **Michel Breger**, michel.breger@univie.ac.at
Editorial Assistant: **Isolde Müller**, isolde.mueller@univie.ac.at
Layout & Production Manager: **Isolde Müller**, isolde.mueller@univie.ac.at

CoAst Editorial and Production Office
Türkenschanzstraße 17, A - 1180 Wien, Austria
http://www.oeaw.ac.at/CoAst/
Comm.Astro@univie.ac.at

Cover Illustration
Sample Phase Distribution Diagram
(For more information see page 36.)

British Library Cataloguing in Publication data.
A Catalogue record for this book is available from the British Library.

ISBN 978-3-7001-7194-2
ISSN 1021-2043

Austrian Academy of Sciences Press
A-1011 Wien, Postfach 471, Postgasse 7/4
Tel. +43-1-515 81/DW 3402-3406, +43-1-512 9050
Fax +43-1-515 81/DW 3400
http://verlag.oeaw.ac.at, e-mail: verlag@oeaw.ac.at

Comm. in Asteroseismology
Volume 163, 2011

Preface

Jeffrey D. Scargle

Space Science and Astrobiology Division, NASA Ames Research Center

SIGSPEC is a method for detecting and characterizing periodic signals in noisy data. This is an extremely common problem, not only in astronomy but in almost every branch of science and engineering. This work will be of great interest to anyone carrying out harmonic analysis employing Fourier techniques.

The method is based on the definition of a quantity called *spectral significance* – a function of Fourier phase and amplitude. Most data analysts are used to exploring only the Fourier amplitude, through the power spectrum, ignoring phase information. The Fourier phase spectrum can be estimated from data, but its interpretation is usually problematic. The spectral significance quantity conveys more information than does the conventional amplitude spectrum alone, and appears to simplify statistical issues as well as the interpretation of phase information.

Peter Reegen
18.8.1968 – 5.2.2011

Comm. in Asteroseismology
Volume 163, 2011

SigSpec User's Manual

P. Reegen†

Institut für Astronomie, Türkenschanzstraße 17, 1180 Vienna, Austria

Abstract

SigSpec computes the spectral significance levels for the DFT* amplitude spectrum of a time series at arbitrarily given sampling. It is based on the analytical solution for the Probability Density Function (PDF) of an amplitude level, including dependencies on frequency and phase and referring to white noise. Using a time series dataset as input, an iterative procedure including step-by-step prewhitening of the most significant signal components and MultiSine least-squares fitting is provided to determine a whole set of signal components, which makes the program a powerful tool for multi-frequency analysis. Instead of the step-by-step prewhitening of the most significant peaks, the program is also able to take into account several steps of the prewhitening sequence simultaneously and check for the combination associated to a minimum residual scatter. This option is designed to overcome the aliasing problem caused by periodic time gaps in the dataset. SigSpec can detect non-sinusoidal periodicities in a dataset by simultaneously taking into account a fundamental frequency plus a set of harmonics. Time-resolved spectral significance analysis using a set of intervals of the time series is supported to investigate the development of eigenfrequencies over the observation time. Furthermore, an extension is available to perform the SigSpec analysis for multiple time series input files at once. In this MultiFile mode, time series may be tagged as target and comparison data. Based on this selection, SigSpec is capable of determining differential

†Note from the Editor: We report with sadness that Peter Reegen, the developer of SigSpec and author of this manual, unexpectedly passed away this year. SigSpec was one of the main achievements of his scientific career. During the previous year he was able to observe the adoption of his program by a large number of astronomers. We sincerely regret the loss of our dear friend and colleague "Piet".
We thank Michael Gruberbauer, who has thoroughly looked through the paper and provided several comments and supplement information. These are represented in uncounted footnotes marked with *.

*Note by M. Gruberbauer: Discrete Fourier Transform (DFT)

significance spectra for the target datasets with respect to coincidences in the comparison spectra. A built-in simulator to generate and superpose a variety of sinusoids and trends as well as different types of noise completes the software package at the present stage of development.

1. What is SigSpec?

SigSpec (abbreviation of '**SIG**nificance **SPEC**trum') is a program that computes a significance spectrum for a time series. It evaluates the *Probability Density Function (PDF)* of a given DFT amplitude level analytically, making use of the theoretical concept introduced by Reegen (2005, 2007). The *False-Alarm Probability*, $\Phi_{\mathrm{FA}}(A)$, is the probability that an amplitude in the DFT spectrum exceeds a given limit A, and is obtained through integration of the PDF (e. g. Scargle 1982). Instead of this frequently used quantity, SigSpec calculates the *spectral significance* (abbreviated by 'sig') of an amplitude A by

$$\mathrm{sig}(A) := -\log\left[\Phi_{\mathrm{FA}}(A)\right] . \tag{1}$$

E. g., a sig equal to 5 indicates that the considered amplitude level is due to noise in one out of 10^5 cases.* This value is used as the default threshold for the termination of the prewhitening sequence.

SigSpec performs an iterative process consisting of four steps[1]:

1. computation of the significance spectrum,
2. exact determination of the peak with maximum sig,
3. a MultiSine least-squares fit of the frequencies, amplitudes and phases of all significant signal components detected so far,
4. prewhitening of the sinusoidal components. The residuals are used as input for the next iteration.

If SigSpec is called without any special settings, it produces four files:

1. the DFT amplitude spectrum `s000000.dat` of the original time series, containing also sig and phase,
2. the DFT amplitude spectrum `resspec.dat` of the residual time series after prewhitening all significant signal components, containing also sig and phase,

*Note by M. Gruberbauer: Note that this quantitiy is defined for a single frequency rather than for the whole spectrum. See Section 15.5. for a discussion.

[1]The AntiAIC computation (p. 47) differs slightly from this procedure.

3. the residual time series `residuals.dat` after prewhitening all significant signal components,

4. a result file called `result.dat`, which contains a list of significant signal components,

5. MultiSine track files, each of which contains a list of the frequencies, amplitudes and phases for a single sinusoidal component through the prewhitening cascade (pp. 36, 87).

Further options may be applied to obtain spectra, residuals, and/or result files (p. 95) in the prewhitening sequence. The MultiSine fits, which are performed after each prewhitening step, modify the frequencies, amplitudes and phases of previous components. If the user examines the resulting signal components and decides not to use all of them, the additional result files help to have accurate frequencies, amplitudes and phases in hands also for a shorter list of significant sinusoids without re-running the program.

SigSpec can produce additional files containing

1. a spectral window for the given time series (pp. 29, 96),

2. a sampling profile (pp. 30, 90) containing the parameters $\alpha_0(\omega)$, $\beta_0(\omega)$, $\theta_0(\omega)$ determining the dependency of the sig on the time-domain sampling, as well as on frequency and phase in Fourier space (see Reegen 2007),

3. a preview of the SigSpec analysis (pp. 39, 90),

4. a Sock Diagram (pp. 31, 94),

5. a Phase Distribution Diagram (pp. 34, 89) containing probability densities for the Fourier phases,

6. a correlogram for each step of the prewhitening sequence (pp. 41, 86).

These options are deactivated by default.

Given a sequence of prewhitenings yielding N significant components with associated sigs $\mathrm{sig}(A_n)$, it is desirable to additionally know the probability of the entire sequence to be valid. This means that not a single erroneous component is allowed. The False-Alarm Probability $\Phi_{\mathrm{FA}\,n} = 10^{-\mathrm{sig}(A_n)}$ of an individual peak is the probability that it is generated by noise. The complementary probability that the considered peak is true is $1 - 10^{-\mathrm{sig}(A_n)}$. If the individual components are statistically independent, the cumulative probability of all components to be real is the product of the individual probabilities,

$$1 - \Phi_{\mathrm{FA}} = \prod_{n=1}^{N} \left(1 - \Phi_{\mathrm{FA}\,n}\right) . \tag{2}$$

Consistently, the cumulative sig is introduced as the negative logarithm of this total False-Alarm Probability for all identified signal components, Φ_{FA}, and in terms of individual sigs, one obtains

$$\mathrm{csig}\left(A_N\right) := -\log\left\{1 - \prod_{n=1}^{N}\left[1 - 10^{-\mathrm{sig}(A_n)}\right]\right\} . \qquad (3)$$

In consistency with the definition of the sig associated with an amplitude in the DFT spectrum, a cumulative sig of 3 means that the prewhitening cascade is entirely true in 999 out of 1 000 cases. Or – in other words – in one out of 1 000 cases, at least one of the identified components is generated by noise.

Whereas the individual sig of a component in the prewhitening sequence may exceed that of the previously identified maximum, the cumulative sig is a monotone sequence uniquely decreasing with each additional signal component.

The prewhitening loop stops, if no sig level above a pre-defined limit is found. As described in "Program termination", p. 23, there are three different criteria that may be applied to determine the conditions for program termination:

1. the number of iterations in the prewhitening sequence,
2. a lower sig limit for the highest peak in the significance spectrum,
3. a threshold for the cumulative sig related to a combined probability for all detected frequency components.

The program also supports the subdivision of a time series into a set of intervals and the separate analysis of all these parts in order to monitor frequency changes of signal components with time. This method will be called *time-resolved analysis*. In this case, the output is somewhat richer, as described in "Time-resolved Analysis" (p. 42).

An immanent problem in the analysis of non-equidistantly sampled time series is *aliasing*. Due to periodic gaps in the data set, a peak in the amplitude spectrum is accompanied by side peaks. Especially if more than one sinusoidal component is present in the data, the superposition of side peaks may produce a maximum amplitude in the DFT spectrum at a frequency that has nothing in common with the true signal frequencies. Such a misidentification usually damages the complete prewhitening sequence from this point on. As pointed out by Reegen (2007), SigSpec appears less prone to aliasing than the previously used methods, since the noise component is employed into the statistical treatment correctly. However, the superposition mentioned above may also lead to erroneous identifications.

In order to overcome this potential weakness, SigSpec supports the simultaneous calculation of more than one signal component simultaneously. Instead of picking only the peak associated to maximum sig, a whole set of highest peaks is examined, searching all possible combinations for several iterations in order to obtain the solution providing a minimum rms residual. This function is called *AntiAIC* (ANTI-ALiasing Correction) mode (p. 47).

There is a second option to examine multiple peaks simultaneously: a non-sinusoidal periodicity is represented by multiple peaks in the DFT amplitude spectrum. One finds a fundamental frequency, plus one or more harmonics the frequencies of which are integer multiples of the fundamental. In astronomical applications, this may occur if shock waves are present in the stellar pulsation or if surface variations are examined. In such a case, it is desirable to take into account not only the fundamental frequency, but also all available harmonics at once. This analysis of harmonics is described on p. 51).

SigSpec is capable of analysing multiple time series input files simultaneously. This MultiFile mode (p. 56) speeds up the computation considerably for time series with the same sampling.

A further option is the evaluation of differential significance spectra (p. 58). The user may specify target vs. comparison data among the input files. Then SigSpec performs a quantitative comparison of the two groups of time series and returns a measure of the probability that a peak in a target dataset is 'true', taking into account amplitudes and phases at the corresponding frequency in the comparison spectra. In this context, the term 'true' is used in the sense of 'not entirely produced by the same variability as present in the comparison data'.

The examples presented here refer to the sample projects available for download at `http://www.SigSpec.org`.

2. How to Run SigSpec

2.1. Projects

SigSpec is called by the command line

```
SigSpec <project>
```

where `<project>` is the name (or path, if desired) of the SigSpec project. Before running the program, the user has to provide

1. a directory `<project>` used for the output,

2. a time series input file (see "The time series input file", p. 12).

The project directory and the time series input file have to be located in the same folder. The project directory need not be empty.

Caution: SIGSPEC **overwrites existing output files!**

There are two conventions for denominating input files.

1. The standard method is to pass only one time series input file to the program. SIGSPEC expects the file to be named `<project>.dat`.

2. For an all-in-one analysis of multiple time series input files, i. e., for running SIGSPEC in MultiFile mode, a leading six-digit index is expected. In this case, the first file shall be named `000000.<project>.dat`, the next file is `000001.<project>.dat`, and so on. For more information on the MultiFile mode, please refer to "MultiFile mode", p. 56.

Furthermore, the user may pass a set of specifications to SIGSPEC by means of a file `<project>.ini` (see "The .ini file", p. 12). This file is expected in the same folder as the time series input file and the project directory. For specifications not given by the user, defaults are used.

Example. *The sample project* `SigSpecNative` *provides a run without any additional options. The command line is* `SigSpec SigSpecNative`. *The sample input file* `SigSpecNative.dat` *(381 data points) represents V magnitudes of IC 4996 # 89 (Zwintz et al. 2004; Zwintz & Weiss 2006).*

The screen output produced by typing `SigSpec SigSpecNative` *at runtime starts with a standard header.*

```
 SSSSSS  ii          SSSSSS
SS    SS            SS    SS
SS       ii  gggg g SS       p pppp   eeeee   ccccc
SS       ii gg   gg SS       pp   pp ee   ee cc   cc
 SSSSSS  ii gg   gg  SSSSSS  pp   pp ee   ee cc
      SS ii gg   gg       SS pp   pp eeeeeee cc
      SS ii gg   gg       SS pp   pp ee      cc
SS    SS ii gg   gg SS    SS pp   pp ee   ee cc   cc
 SSSSSS  ii  gggggg  SSSSSS  pppppp   eeeee   ccccc
                 gg          pp
            gg   gg          pp
             ggggg           pp


SIGnificance SPECtrum
Version 2.0
****************************************************************
by Piet Reegen
Institute of Astronomy
University of Vienna
Tuerkenschanzstrasse 17
1180 Vienna, Austria
Release date: August 18, 2009
```

SIGSPEC *processes the command line, checks whether a project directory* `SigSpecNative` *is present, and searches for a file* `SigSpecNative.ini` *(see*

"The `.ini` file", p. 12). Since there is no such file present, four warning messages are produced.

```
*** start ************************************************

command line interface
Checking availability of project directory SigSpecNative...
project directory SigSpecNative ok.
loading .ini file

Warning: IniFile_SSCols 001
         Failed to open .ini file.


Warning: IniFile_WCols 001
         Failed to open .ini file.


Warning: IniFile_LoadIni 001
         Failed to open .ini file.


Warning: IniFile_Cind 001
         Failed to open .ini file.
```

The next task is to load the input file `SigSpecNative.dat`. SIGSPEC *provides the number of rows, the time interval width, and the standard deviation of the observable.*

```
*** loading time series input file(s) *********************

SigSpecNative.dat

*** time series properties ********************************

points 381, time base 9.17532, rms dev 0.00449592
```

The next section contains the specifications for the DFT and significance spectra to be calculated.

```
*** preparing to run SigSpec ******************************

Rayleigh frequency resolution            0.1089880382935977
oversampling ratio                      20.0000000000000000
frequency spacing                        0.0054494019146799
lower frequency limit                    0.0000000000000000
upper frequency limit                  100.4651736990383739
Nyquist coefficient                      0.5000000000000000
number of frequencies               18437
```

As SIGSPEC *performs the prewhitening sequence, a list of detected signal components is displayed. The screen output contains the index of the identified component (a line number), the sig, the time-domain rms deviation before prewhitening the corresponding signal, and the csig. The last line contains an insignificant component that meets the breakup condition. In the present example, the default breakup condition (the sig to drop below 5) is applied, which is satisfied in the fourth iteration, where the maximum sig is 4.10802.*

```
*** running SigSpec ****************************************

  1 freq 3.13205  sig 9.54539  rms 0.00449592  csig 9.54539
  2 freq 3.98471  sig 7.43085  rms 0.00422861  csig 7.42753
  3 freq 5.40684  sig 5.30164  rms 0.0040257  csig 5.2984
  4 freq 17.3677  sig 4.13698  rms 0.00388775  csig 4.10802
```

On exit, SIGSPEC *displays a good-bye message.*

```
Finished.

************************************************************

Thank you for using SigSpec!
Questions or comments?
Please contact Piet Reegen (reegen@astro.univie.ac.at)
Bye!
```

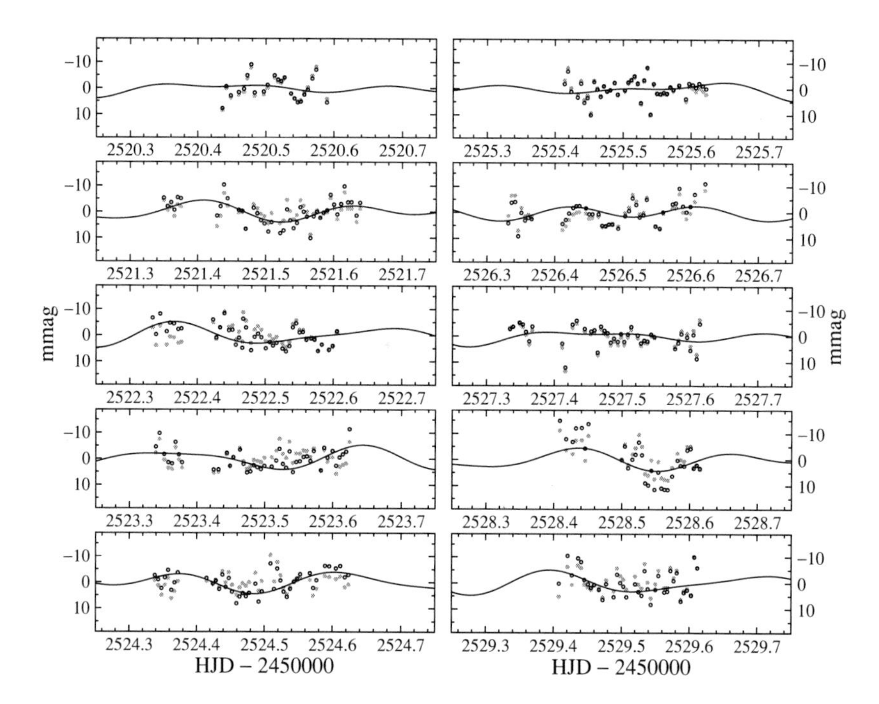

Figure 1: *Black circles:* light curve for the sample project SigSpecNative. *Line:* fit formed by three significant signal components (as listed in the file SigSpecNative/result.dat). *Grey dots:* residuals after prewhitening of three significant signal components (file SigSpecNative/residuals.dat).

If no special output is selected in a file SigSpecNative.ini, SIGSPEC *produces the following output files in the project directory* SigSpecNative*:*

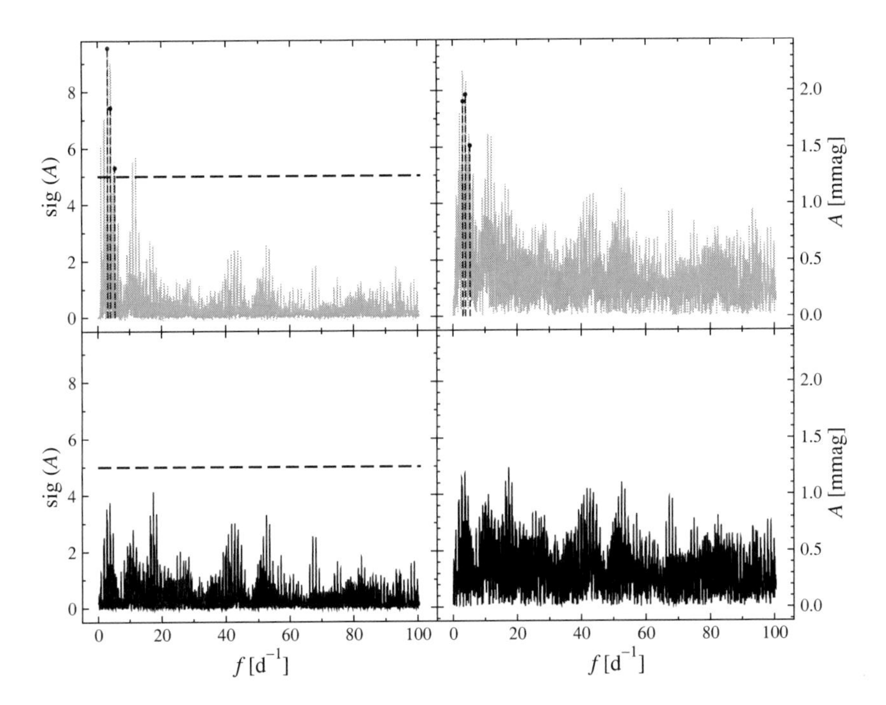

Figure 2: *Grey:* Fourier spectra for the sample project `SigSpecNative`. *Left:* significance spectra. *Right:* DFT amplitudes. *Top:* original spectra, without prewhitening (file `SigSpecNative/s000000.dat`). *Bottom:* residual spectra, with three significant signal components prewhitened (file `SigSpecNative/resspec.dat`). In the top panels, the significant components are indicated by *dots* with dashed drop lines (file `SigSpecNative/result.dat`). The default sig threshold of 5 is represented by a horizontal *dashed line* in the left panels.

- `s000000.dat`*: DFT and significance spectrum of the original time series (without any prewhitening),*
- `result.dat`*: list of significant signal components detected in the time series,*
- `residuals.dat`*: residual time series after prewhitening all significant signal components listed in* `result.dat`*,*
- `resspec.dat`*: DFT and significance spectrum of the residual time series* `residuals.dat`*.*

Fig. 1 contains the sample input `SigSpecNative.dat`*, the multisine fit to the time series according to the list of significant signal components in*

`SigSpecNative/result.dat`, *and the residuals after subtracting the fit from the input time series (file* `SigSpecNative/residuals.dat`*). Fig. 2 refers to the frequency domain: the DFT spectrum of the initial time series* `SigSpecNative/s000000.dat`, *the three significant signal components contained in* `SigSpecNative/result.dat`, *and the residual spectrum in the file* `SigSpecNative/resspec.dat`. *For detailed information on the contents of the output files, please refer to "Default Output", p. 24.*

Furthermore, the user may pass a set of specifications to SigSpec in a file `<project>.ini` (see "The `.ini` file", p. 12). For specifications not given by the user, defaults are used.

2.2. Quiet mode

If the command line is followed by the letter 'q', i. e.

```
SigSpec <project> q
```

all screen output is suppressed.

3. Input

3.1. The time series input file

The input file for SigSpec is a time series. The corresponding file has to be located in the same folder as the project directory. The only restrictions to the format are that the number of items per row has to be constant for all rows in the file and that columns have to be separated by at least one whitespace character or tab.* Dataset entries need not be numeric, except for the columns specified as time, observable, and weights (p. 13).

3.2. The `.ini` file

An optional file `<project>.ini` consists of a set of keywords and arguments defining project-specific parameters for SigSpec. If this file is not present in the same folder as the time series input file(s), SigSpec uses a set of default parameters. A complete list of keywords is given in "Keywords Reference", p. 85.

Multiple use of the same keyword or the specification of contradictory keywords causes the software to take into account only the last declaration. There are only three exceptions:

*Note by M. Gruberbauer: Caution: SigSpec does not support the exponential annotation (e. g. `4.234E03` or `1.0385e-03`)!

1. SigSpec accepts multiple weights columns specified by `col:weights` (p. 13),
2. multiple subset identifier columns may be specified by `col:ssid` (p. 15),
3. the simulator may be used to synthesize multiperiodic signal plus various types of noise upon the given sampling, where the keywords `sim:signal`, `sim:poly`, `sim:exp`, `sim:zeromean`, `sim:serial`, `sim:temporal`, and `sim:rndsteps` may be used multiply (see "The simulator mode", p. 63).

Caution: SigSpec expects a carriage-return character at the end of the file `<project>.ini`, otherwise the program may hang!

Lines in the `.ini` file starting with a # character are ignored by SigSpec. This provides the possibility to write comments into the file. Furthermore, additional characters beyond what is expected in a line (keyword plus required number of parameters) is ignored. Thus it is allowed to add comments also at the end of the lines containing relevant information for SigSpec.

3.3. Time series columns representing time and observable

The keywords `col:time` and `col:obs` determine those columns in the time series input file which contain time values and the observable monitored over time, respectively. These columns are required and have to be uniquely specified. Column indices start with 1.

If `col:time` is not specified, the default value 1 is used. If `col:obs` is not specified, the default value 2 is used.

Example. *The sample project* `coltimecolobs` *contains a dataset where the time and observable values are found in columns 2 and 3, respectively. The input time series represents the V photometry of IC 4996 # 89 (see Example* `SigSpecNative`*, p. 8). The file* `coltimecolobs.ini` *contains the two lines*

```
col:time 2
col:obs 3
```

3.4. Time series columns containing statistical weights

Furthermore, one or more columns with statistical weights may be specified using the keyword `col:weights`. The keyword accepts two arguments: the first is the column index, the second is a floating-point value, say p_n for the nth weights column. Given N weights columns indexed according to $n = 1, \ldots, N$, the total weight for the mth row is evaluated using the weights w_{nm} in the individual columns according to

$$\Gamma_m := \prod_{n=1}^{N} \gamma_{nm}^{p_n} . \tag{4}$$

Weights need not be normalised, this is performed by SigSpec.

Time, observable, and weights columns have to consist of floating-point numbers only. SigSpec checks these columns before starting the computations. If a non-numeric entry is found in one of the special columns, the program terminates with an error message.

Caution: SigSpec does not support the exponential annotation (e. g. `4.234E03` or `1.0385e-03`)!

Example. *The sample project* `weights` *contains a dataset with statistical weights in column 3 the squares of which are used by SigSpec, as specified by the* `.ini` *file entry*

```
col:weights 3 2
```

The input time series `weights.dat` *represents the sampling of IC 4996 # 89 (V), and the magnitudes were synthesized by*

1. *a sinusoid with frequency 4.68573 cycles per day, amplitude 17.27 mmag,*
2. *Gaussian noise with 25 mmag rms deviation only for the measurements between HJD 2452524 and HJD 2452525,*
3. *Gaussian noise with 2.5 mmag rms deviation for all other nights.*

The resulting light curve is displayed in Fig. 3. Fig. 4 compares the frequency domain output (a closeup for frequencies below 10 cycles per day) with and without employing the weights. Without weights, the peak at 4.7 cycles per day visible, but not the most significant one. Moreover, there is no signal that reaches the sig threshold of 5.[2]

```
1 freq 5.68136  sig 3.75547  rms 0.088716  csig 3.75547
```

Column 3 in the time series input file `weights.dat` *contains zeroes for the measurements between HJD 2452524 and HJD 2452525 and values of 1 for the rest. Consequently, in this example, the exponent 2 assigned to the keyword* `col:weights` *in the file* `weights.ini` *does not affect the weighting: the result would be the same if, e. g.,*

[2] The result without weights is found in the project directory `noweights`.

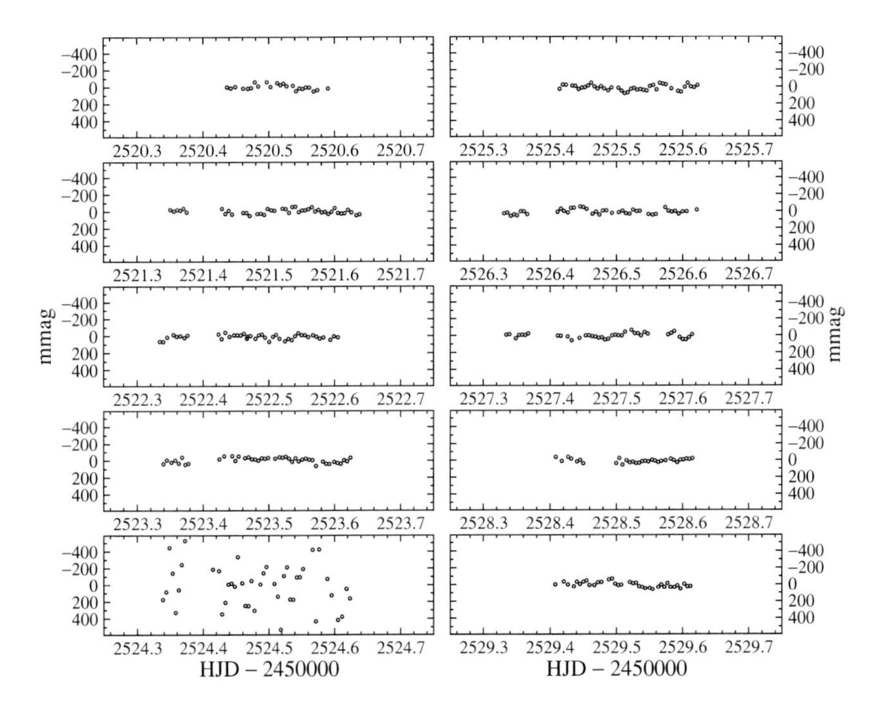

Figure 3: Light curve for the sample project `weights`.

```
col:weights 3 1
```

were given instead of

```
col:weights 3 2
```

Employing the weights column, SigSpec *provides the component at 4.7 cycles per day as the only significant signal:*

```
1 freq 4.67968  sig 20.395   rms 0.029129   csig 20.395
2 freq 30.5489  sig 4.47468  rms 0.0252866  csig 4.47468
```

3.5. Time series columns containing subset identifiers

If the mean magnitude of a light curve is desired to be adjusted to zero for each night, or if the data are obtained from more than one site, one may perform an individual zero-mean correction for subsets of the total time series. This is achieved by the keyword `col:ssid` in the .ini file. This keyword is followed by the index of the column which shall be assigned to subset identifiers and may

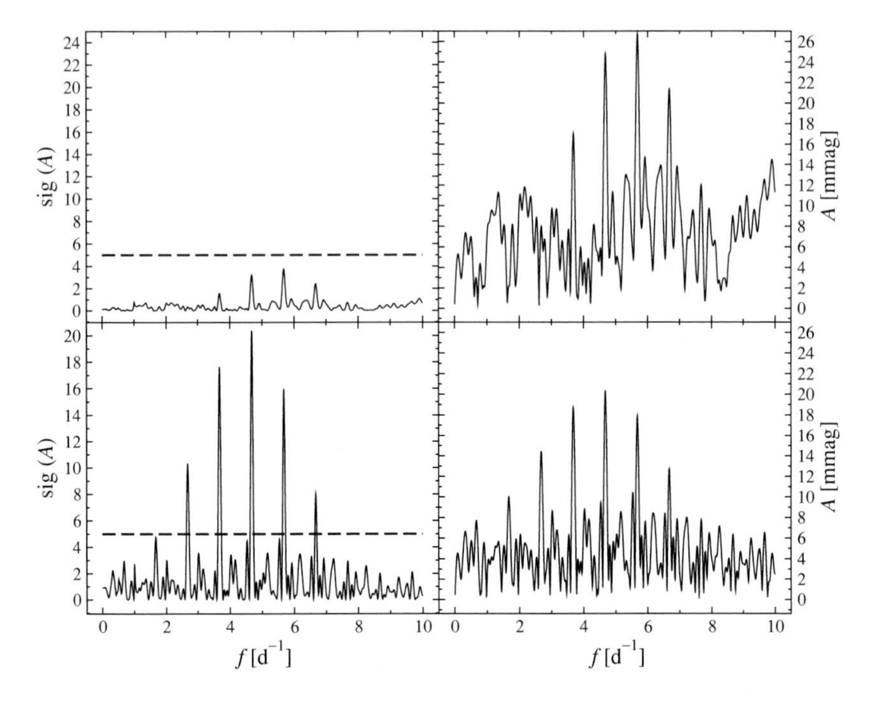

Figure 4: *Grey:* Fourier spectra for the sample project `weights`. *Left:* significance spectra. *Right:* DFT amplitudes. *Top:* spectra of the unweighted time series. *Bottom:* spectra employing statistical weights. The significant components are indicated by *dots* with dashed drop lines (file `weights/result.dat`). The default sig threshold of 5 is represented by a horizontal *dashed line* in the left panels.

be multiply defined, if more than one subset identifier column is given. Subset identifiers may be arbitrary alpha-numeric strings.

If `col:ssid` is specified, SigSpec treats all lines in the dataset with equal subset identifiers as individual subsets, for each of which a zero-mean correction is performed. Subsequently, SigSpec performs the appropriate statistical calculations, taking into account that the zero-mean correction for subsets diminishes the degrees of freedom for noise.

If more than one subset column is specified, data points are considered to belong to the same subset, if all corresponding subset identifiers are equal.

Example. *The sample project* `subsets` *contains a dataset with subset identifiers in column 3. The input time series* `subsets.dat` *represents the sampling of IC 4996 # 89 (V), and the magnitudes were synthesized by adding*

1. *Gaussian noise with 5 mmag rms deviation,*

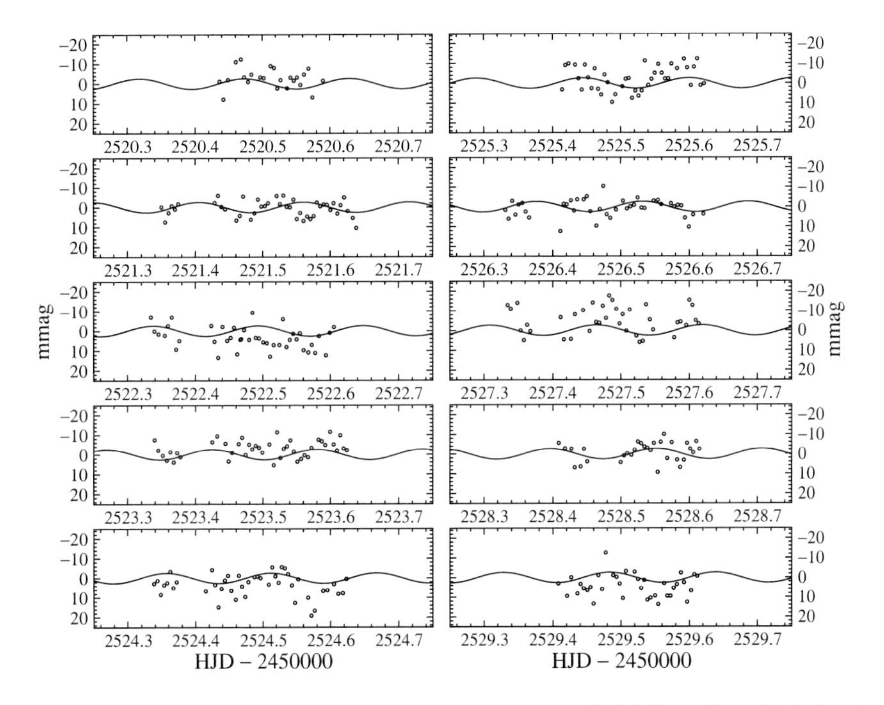

Figure 5: Light curve for the sample project `subsets`. *Solid line:* Sinusoidal signal used as input.

2. *a sinusoid with frequency 6.43682 cycles per day and amplitude 2.62 mmag,*

3. *individual constant zeropoint offsets on a millimag range for each night.*

The resulting light curve is displayed in Fig. 5, displaying the input signal as a solid line and the data points including the nightly offsets as open dots. Fig. 6 compares the resulting frequency domain output (a closeup for frequencies below 10 cycles per day) with and without employing the weights. If no subdivision according to the subset identifiers is performed, the spectra show a peak at 6.4 cycles per day plus several spurious peaks at frequencies below 2 cycles per day, which are due to the interpretation of the nightly shifts as signal in the 1-cycle-per-day domain and also harmonics.[3] *Consequently,* SigSpec *identifies two additional significant signal components at low frequencies:*

[3]The result without subsets is found in the project directory `nosubsets`.

```
1 freq 0.575256  sig 14.5784  rms 6.0813  csig 14.5784
2 freq 7.43176  sig 6.39232  rms 5.51956  csig 6.39232
3 freq 0.286066  sig 5.21585  rms 5.29531  csig 5.18785
4 freq 75.1664  sig 3.39587  rms 5.12278  csig 3.38892
```

Column 3 in the time series input file `weights.dat` *contains characters A to J uniquely assigned to each night. Employing the subsets column eliminates the low-frequency artefacts, and* SigSpec *provides the component at 6.4 cycles per day as the only significant signal:*

```
1 freq 6.4376  sig 8.20485  rms 5.2253  csig 8.20485
2 freq 75.1661  sig 3.70924  rms 4.95954  csig 3.70922
```

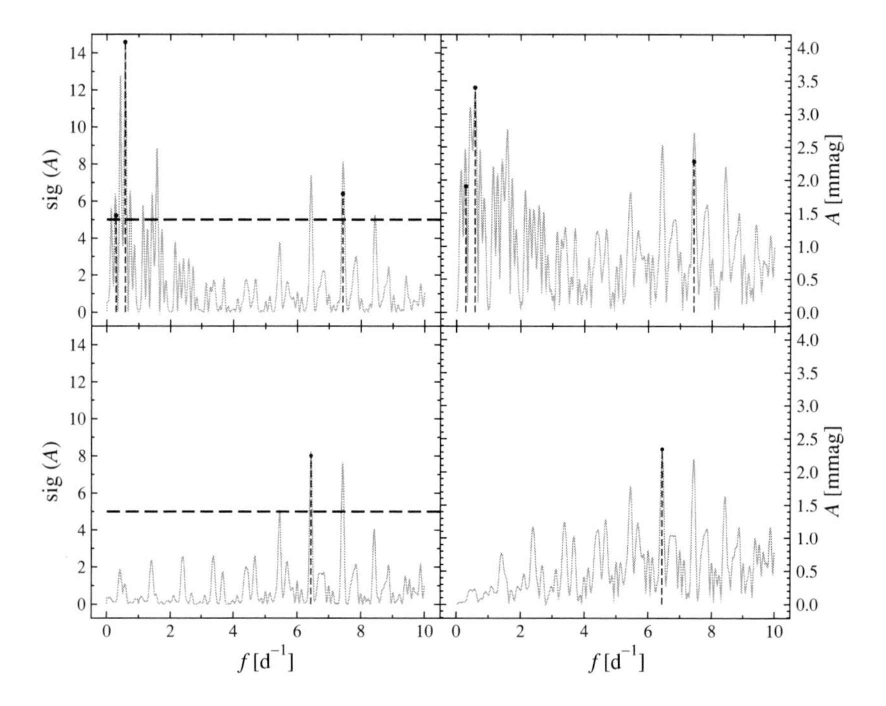

Figure 6: *Grey:* Fourier spectra for the sample project `subsets`. *Left:* significance spectra. *Right:* DFT amplitudes. *Top:* spectra of the total time series. *Bottom:* spectra of the subdivided time series. The significant components are indicated by *dots* with dashed drop lines (file `subsets/result.dat`). The default sig threshold of 5 is represented by a horizontal *dashed line* in the left panels.

3.6. Lower frequency limit

The frequency where the computation of spectra starts is specified by the keyword `lfreq`. By default, the lower frequency limit is zero.

Example. *The sample project* `limits` *illustrates the use of the keyword* `lfreq`. *It uses the V photometry of IC 4996 # 89 as input file* `limits.dat`, *and the file* `limits.ini` *contains the line*

```
lfreq 1
```

which forces SIGSPEC *to perform all computations for frequencies* $\geq$ *1 cycle per day. The spectrum* `limits/s000000.dat` *is displayed in Fig. 7.*

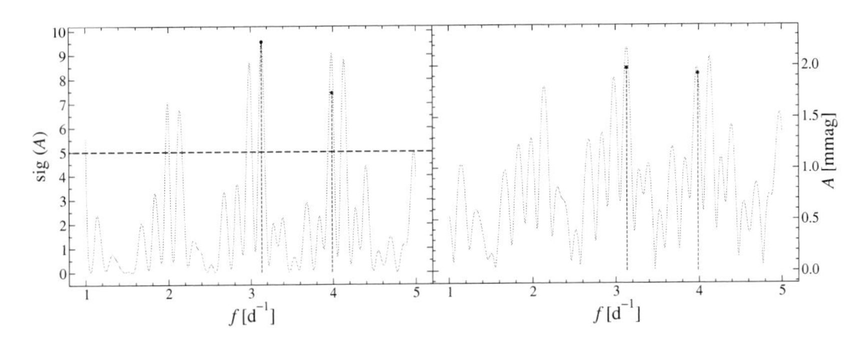

Figure 7: *Grey:* Fourier spectra for the sample project `limits`. *Left:* significance spectrum. *Right:* DFT amplitudes. The significant components are indicated by *dots* with dashed drop lines (file `limits/result.dat`). The default sig threshold of 5 is represented by a horizontal *dashed line* in the left panel. The frequency range is set from 1 to 5 cycles per day using the keywords `lfreq` and `ufreq`.

3.7. Upper frequency limit and Nyquist Coefficient

The keyword `ufreq` allows to determine the upper limit of the frequency interval to be considered.

An alternative method is the automatic determination of this limit by means of the Nyquist Coefficient (keyword `nycoef`). For equidistantly sampled time series with sampling interval width δt, there is a uniquely defined *Nyquist Frequency*

$$f_\nu := \frac{1}{2\,\delta t} \,. \tag{5}$$

In case of non-equidistant sampling, each sampling interval between two consecutive time values may be considered to produce its individual Nyquist Frequency, whence this limit is ambiguous. In this case, the Nyquist Coefficient for an arbitrarily given frequency is introduced as the fraction of sampling intervals in the time domain the individual Nyquist Frequency of which is higher than the frequency under consideration. This provides to select an upper frequency limit by specifying a minimum Nyquist Coefficient. E. g., specifying a Nyquist Coefficient of 0.5 (which is the default value) guarantees that at least half of the information contained by the spectrum in the considered frequency range is below the Nyquist limit.

Additional information is available by setting the keyword `nyscan` in the `.ini` file. If this keyword is specified, SigSpec creates a file `nyscan.dat` in the project directory containing the Nyquist Coefficients over the specified frequency range.

Example. *The sample project* `limits` *illustrates the use of the keyword* `ufreq`. *The line*

```
ufreq 5
```

in the file `limits.ini` *restricts all computations performed by* SigSpec *to frequencies below 5 cycles per day. The spectrum* `limits/s000000.dat` *(sig and amplitude) is displayed in Fig. 7. A comparison with the screen output in Example* `SigSpecNative`, *p. 9, where no restrictions to the frequency range apply, shows that the screen output in this example contains one line less:*

```
1 freq 3.13205  sig 9.54539  rms 0.00449592  csig 9.54539
2 freq 3.98471  sig 7.43085  rms 0.00422861  csig 7.42753
3 freq 2.664  sig 4.60182  rms 0.0040257  csig 4.60117
```

The signal component at 5.4 cycles per day is not detected, because it is outside the specified frequency range.

Example. *The sample project* `nyos` *illustrates the use of the keywords* `nycoef` *and* `nyscan` *for the V photometry of IC 4996 # 89. The line*

```
nycoef 0.99
```

in the file `nyos.ini` *provides an upper frequency limit of 110.77 cycles per day. The keyword* `nyscan` *is given, and the file* `nyos/nyscan.dat` *contains the Nyquist Coefficients for frequencies from 0 to 110.77 cycles per day, as displayed in Fig. 8.*

3.8. Frequency spacing and oversampling ratio

The width of the interval between consecutive frequencies may be specified by the keyword `freqspacing`.

An alternative method is the automatic determination of the spacing by means of the oversampling ratio. In case of equidistantly sampled time series, the frequency spacing is defined by

$$\delta f := \frac{1}{T}, \tag{6}$$

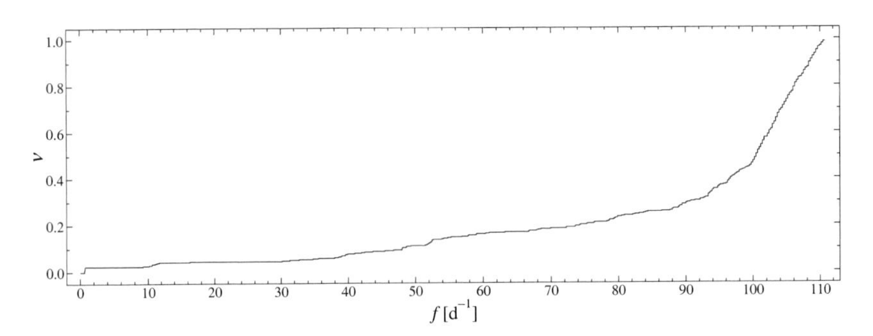

Figure 8: The file `nyos/nyscan.dat` contains the Nyquist coefficients depending on frequency for the sample project `nyos`.

where T denotes the width of the total time interval. For non-equidistant time series, it is advisable to use a denser frequency sampling,

$$\delta f := \frac{1}{\Omega T}, \tag{7}$$

where Ω is the oversampling ratio. This quantity may be specified using the keyword `osratio`. The default value is 20, which is – in most cases – sufficient for practical use.

Example. *The sample project* `limits` *illustrates the use of the keyword* `freqspacing`*, an example for the keyword* `osratio` *is provided in the sample project* `nyos`*. Both samples use the V photometry of IC 4996 # 89 as input time series. The line*

```
freqspacing 0.001
```

in the file `limits.ini` *forces* SIGSPEC *to calculate Fourier amplitudes and sigs for every 0.001 cycles per day. In the file* `nyos.ini`*, there is a line*

```
osratio 12
```

which overrides the default oversampling ratio of 20. Fig. 9 compares the standard spacing from Example `SigSpecNative`*, p. 8), with the spacings obtained applying the two above modifications.*

3.9. Accuracy of MultiSine fits

By default, SIGSPEC performs a MultiSine least-squares fit each time a new significant signal component is detected. The fitting procedure improves the frequencies, amplitudes, and phases of all previously detected signal components.

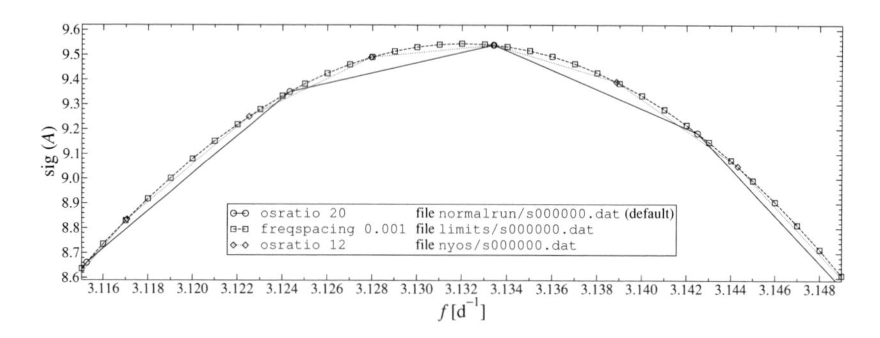

Figure 9: Close-up for the significance spectra generated by the projects `SigSpecNative`, `limits` and `nyos` around the main peak for the V photometry of IC 4996 # 89. Different settings for frequency spacing and oversampling ratio are applied.

The algorithm applies Newton's root finding scheme to the first derivatives of the residual variance.

The precision of computed frequencies via MultiSine least-sqares fits is defined according to

$$\delta f := \frac{\mu}{T \,\mathrm{sig}^{\frac{\kappa}{2}}} , \tag{8}$$

where μ and κ are the accuracy parameters for MultiSine fitting. The default value of μ is 10^{-6}, that of κ is 1. They may be adjusted by the keyword `multisine:newton`, followed by μ, κ and a third parameter determining the relative tolerance of the time-domain rms error between consecutive iterations (see next paragraph). To reduce the potential time consumption of the procedure, μ can be adjusted to achieve an overall scaling of the frequency accuracy. The value of κ determines the dependence of the demanded frequency precision on the sig of the peak under consideration. Setting $\kappa = 0$ yields the Rayleigh frequency resolution, for $\kappa = 1$ one obtains the Kallinger resolution (Kallinger, Reegen & Weiss 2008).

The criterion on which MultiSine fitting is based is the minimisation of rms residual. Thus the rms residual is demanded to decrease from one iteration to the next. Otherwise the fitting procedure is terminated. To speed up the computation, the MultiSine fit can be terminated, if the relative improvement of rms residual drops below a positive number. The default value 10^{-6}. This value may be re-adjusted by the third parameter to the keyword `multisine:newton`.

The two termination conditions are linked by a logical 'and', i. e. the MultiSine fitting procedure stops if both the desired frequency accuracy is reached for all signal components and the improvement of residual rms drops below the specified threshold.

There is an optional keyword, `multisine:lock`, that forces the prewhitening cascade to rely on the "raw" frequencies, amplitudes and phases (i. e. those without MultiSine fitting). Resulting signal components are improved to obtain a least-squares fit in each iteration, but this improvement is ignored in the prewhitening sequence. The default setting is that the improved parameters are used for the subsequent analysis (as also obtained by the keyword `multisine:unlock`.

Example. *The sample project* `multisine` *illustrates the application of the keyword* `multisine:newton` *to the IC 4996 # 89 photometry (V) as input file* `multisine.dat`. *The file* `multisine.ini` *contains the line*

```
multisine 0.001 0 0.01
```

which reduces the accuracy of the MultiSine fit, compared to the default values 0.000001, 1, 0.000001, respectively. The second parameter refers to the Rayleigh frequency resolution rather than the (default) Kallinger frequency resolution. The screen output provides four entries:

```
1 freq 3.13205  sig 9.54539  rms 0.00449592  csig 9.54539
2 freq 3.98472  sig 7.43087  rms 0.00422861  csig 7.42755
3 freq 5.40686  sig 5.29838  rms 0.0040257  csig 5.29516
4 freq 17.3677  sig 4.13727  rms 0.00388775  csig 4.10809
```

For comparison, the project `SigSpecNative`, *p. 8, employs the default settings.*

For the first entry, there is no difference between the two results, but due to propagation of uncertainties, the following entries show slight and increasing deviations. As expected, the rms errors of residuals are higher if the accuracy is reduced.

3.10. Program termination

There are three possibilities to specify a criterion for program termination:

1. the number of iterations (keyword `iterations`),
2. a lower sig limit (keyword `siglimit`),
3. the reliability of the entire analysis is determined by the *cumulative sig*. It is based on the probability that at least one of the frequency components detected so far is due to noise. A threshold in terms of cumulative sig may be defined using the keyword `csiglimit` For an introduction to the cumulative sig, please refer to p. 5.

Multiple specifications in terms of these keywords cause the prewhitening cascade to terminate if one of the limits is reached.

The default assignment for `siglimit` is 5. This pre-definition may be deactivated by defining

```
siglimit 0
```

in the `.ini` file. The limits `iterations` and `csiglimit` are switched off by default.

Example. *The sample project* `terminate` *contains a combination of the keywords* `siglimit`, `csiglimit` *and* `iterations`, *applied to the V photometry of IC 4996 # 89 as input file. For a comparison to the standard output, please refer to Example* `SigSpecNative`, *p. 8. The file* `terminate.ini` *contains a combination of three keywords:*

```
siglimit 0
csiglimit 3
iterations 10
```

The first line deactivates the default setting of 5 for the sig limit. The combination of the second and third line forces SigSpec *to terminate after 10 iterations, or earlier, if the cumulative sig drops below 3. The screen output provides seven lines, corresponding to six significant signal components:*

```
1 freq 3.13205  sig 9.54539  rms 0.00449592  csig 9.54539
2 freq 3.98471  sig 7.43085  rms 0.00422861  csig 7.42753
3 freq 5.40684  sig 5.30164  rms 0.0040257  csig 5.2984
4 freq 17.3677  sig 4.13698  rms 0.00388775  csig 4.10802
5 freq 3.67101  sig 3.73187  rms 0.00378701  csig 3.57943
6 freq 52.5182  sig 3.41319  rms 0.00369756  csig 3.18744
7 freq 41.7372  sig 3.02872  rms 0.00361981  csig 2.80001
```

The cumulative sig of 2.8 for component 7 is responsible for program termination before the limit of 10 iterations is reached.

4. Default Output

All output files are written into the project directory. A six-digit index denotes the iteration in the prewhitening cascade. E. g., an index 000000 represents a file obtained from the input data without any prewhitening, 000001 denotes a file after prewhitening of the first sinusoidal component. The general annotation `#iteration#` will be used for this six-digit identifier.

Example.[4] *The sample project* `output` *illustrates how to adjust the output of* SigSpec. *The input file* `output.dat` *represents 16 nights (992 data points) of Strømgren y photometry (Vienna University APT, T6; Strassmeier et al. 1997) of the Delta Scuti star EE Cam (Breger, Rucinski & Reegen 2007). The light curve is displayed in Fig. 10.*

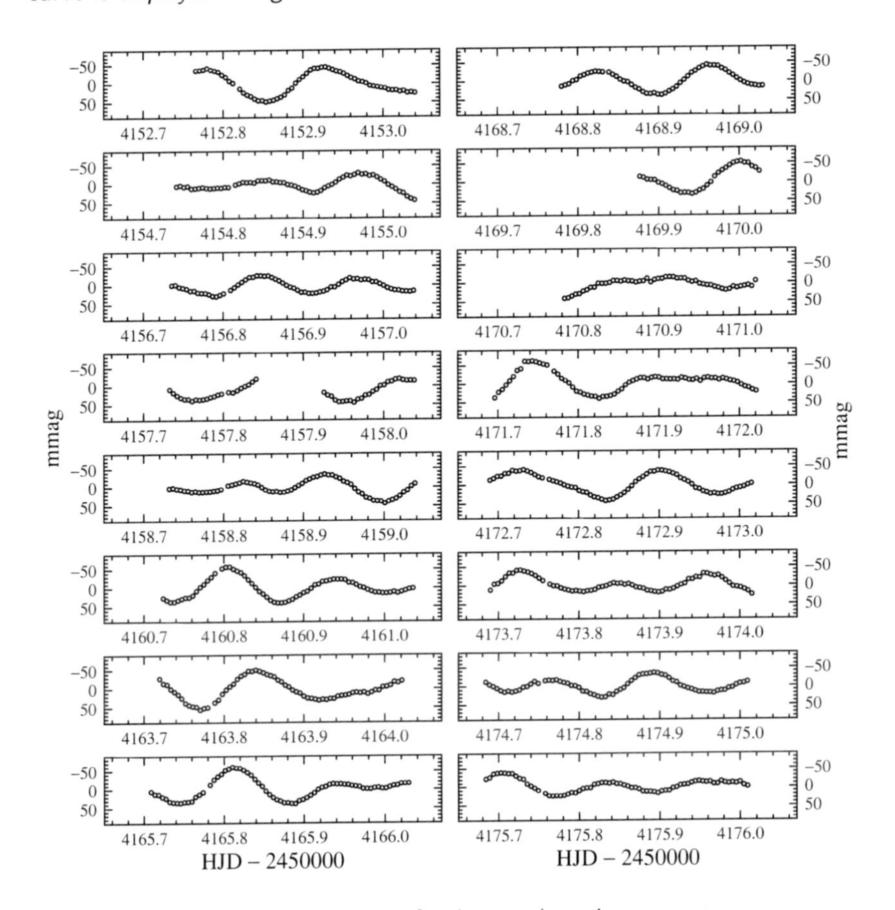

Figure 10: Light curve for the sample project `output`.

[4]The sample project `output` is the most time consuming sample of all. The computation takes 90 minutes on an Intel Core2 CPU T5500 (1.66GHz) under Linux 2.6.18.8-0.9-default i686. This is mostly due to the calculations of the Sock and Phase Distribution Diagrams. In order to speed up the program, the user may switch off these operations by placing a # character at the beginning of all lines containing keywords `sock:...` and `phdist:...` in the file `output.ini`.

The vast amount of output provided by Sock Diagrams and Phase Distribution Diagrams makes it necessary to restrict the frequency interval in the file `output.ini`. *Especially close to zero frequency, the output may be tremendous. Thus the very low frequencies are avoided:*

```
lfreq 1
ufreq 16
```

The frequency spacing is adjusted to speed up the computations of Sock and Phase Distribution Diagrams.

```
freqspacing 0.005
```

All other entries in the file `output.ini` *apply to output files and are discussed in the subsequent sections.*

4.1. Spectra

By default, two spectra (files `s000000.dat` and `resspec.dat`) are generated. The file `s000000.dat` contains the spectrum of the original time series, and the file `resspec.dat` represents the residual spectrum after finishing the prewhitening sequence.

The columns are

1. frequency [inverse time units],
2. sig,
3. DFT amplitude [units of observable],
4. Fourier-space phase angle [rad],
5. Fourier-space phase angle of maximum sig [rad].

To achieve consistency with the output for differential significance spectra (see p. 58), two further columns are found containing values -1 and 0 only.

The phase angles θ are given according to a trigonometric fit,

$$F(t) := A\cos(2\pi ft - \theta) , \tag{9}$$

with amplitude A and frequency f as given in the file. This convention is compatible to the definition of phase in Fourier space. This definition of phase is consistently used for all types of SigSpec output.

If the keyword `spectra` is provided in the `.ini` file, additional output files `s#iteration#.dat` are generated. The index `#iteration#` starts with 000001, denoting the residual spectrum after the first prewhitening step.

The keyword `spectra` expects two integer parameters. The first defines the number of iterations for which these files shall be generated. A negative number causes SIGSPEC to generate files for all iterations. The second parameter has to be a positive number and defines a step width. If it is set 1, a file is generated after each iteration, if it is set 2, after every second iteration (starting with `s000002.dat`), and so on.

Example. *The sample project* `output` *uses the keyword* `spectra` *in the file* `output.ini`, *namely*

```
spectra 10 2
```

Spectra are written only during the first 10 iterations of the prewhitening sequence. The second parameter provides only every second file to be generated. In this example, the following files are produced:

```
output/s000000.dat
output/s000002.dat
output/s000004.dat
output/s000006.dat
output/s000008.dat
output/s000010.dat
```

In addition, the file `resspec.dat` *contains the residual spectrum after all iterations.*

4.2. Residual time series

By default, a file `residuals.dat` is generated. It represents the residuals after prewhitening all signal components found significant. The column format is the same as for the time series input file.

If the keyword `residuals` is provided in the `.ini` file, additional files `t#iteration#.dat` are generated, where the index `#iteration#` starts with 000001, denoting the residuals after the first prewhitening step.

The keyword `residuals` expects two integer parameters. The first defines the number of iterations for which these files shall be generated. A negative number causes SIGSPEC to generate files for all iterations. The second parameter has to be a positive number and defines a step width. If it is set 1, a file is generated after each iteration, if it is set 2, after every second iteration (starting with `t000002.dat`), and so on.

Example. *The sample project* `output` *uses the keyword* `residuals` *in the file* `output.ini`, *namely*

```
residuals -1 5
```

Setting the first parameter -1 provides residual time series during the entire prewhitening sequence. The second parameter provides only fifth second file to be generated. Since the number of iterations in this example is 40, the following files are produced:

```
output/t000005.dat
output/t000010.dat
output/t000015.dat
output/t000020.dat
output/t000025.dat
output/t000030.dat
```

In addition, the file `residuals.dat` *contains the residual time series after all iterations.*

4.3. Result files

The file `result.dat` contains a list of all identified sig maxima. This file consists of seven columns providing

1. frequency [inverse time units],
2. sig,
3. amplitude [units of observable],
4. phase [rad],
5. rms scatter of the time series before prewhitening,
6. point-to-point scatter of the time series before prewhitening,
7. the cumulative sig for all frequency components detected so far.

Columns 3 and 4 represent amplitude and phase as the result of a least-squares fit to the time series at the present prewhitening stage (i. e. after subtraction of all previously identified signal components) for the frequency of maximum significance.

The last line in the file contains zeroes for frequency, amplitude, and phase. The non-zero values refer to the (cumulative) sig of the most significant component below the threshold, and to the rms and point-to-point scatter after the last prewhitening step, respectively. This final line is suppressed if the criterion `iterations` is responsible for program termination.

If the keyword `results` is provided in the `.ini` file, additional result files `r#iteration#.dat` are generated, where the index `#iteration#` starts with 000001, denoting the result of the first iteration. The files contain the significant components within the prewhitening cascade as preliminary results. The

MultiSine least-squares fits which are performed at each step of the prewhitening sequence modify frequencies, amplitudes and phases. Therefore it may be useful to have additional results from earlier iterations in hands, if the user decides not to use all components found by SigSpec without re-running the program.

The keyword `results` expects two integer parameters. The first defines the number of iterations for which these files shall be generated. A negative number causes SigSpec to generate files for all iterations. The second parameter has to be a positive number and defines a step width. If it is set 1, a result file is generated after each iteration, if it is set 2, after every second iteration (starting with `r000002.dat`), and so on.

Example. *The sample project* `output` *uses the keyword* `results` *in the file* `output.ini`*, namely*

```
results -1 1
```

providing result files `r000001.dat`, `r000002.dat`*,..., for all iterations of the entire prewhitening sequence. In addition, the final result after all prewhitening iterations is contained in the file* `results.dat`.

5. Analysis of the Time-domain Sampling

Example. *The sample project* `output` *contains the output of a spectral window, a sampling profile, a sock diagram, a phase distribution diagram, a preview, and correlograms.*

5.1. Spectral window

The spectral window is computed, if the keyword `win` is given in the `.ini` file. This keyword does not require any parameters. The output is provided in the file `win.dat`. It consists of three columns referring to

1. frequency [inverse time units],
2. amplitude [units of observable],
3. Fourier-space phase angle [rad].

Example. *The sample project* `output` *contains the output of a spectral window. The file* `output.ini` *contains the keyword* `win`*, and the corresponding output is found in the file* `output/win.dat` *and displayed in Fig. 11. The frequency limits determined by the lines*

```
lfreq 1
ufreq 16
```

also apply to the spectral window.

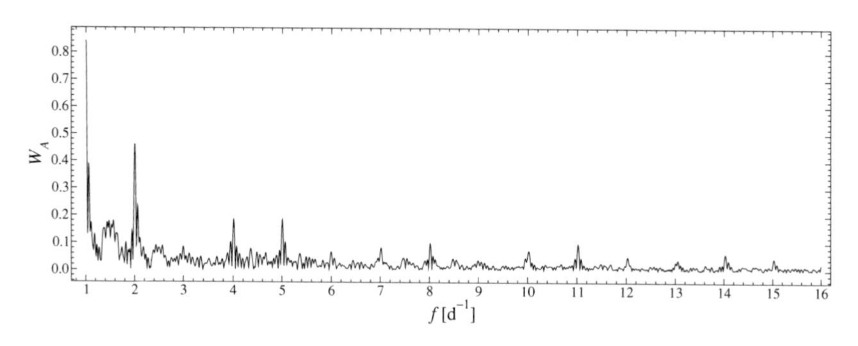

Figure 11: Spectral window for the sample project `output`.

5.2. Sampling profile

The sampling profile is an essential part of the sig computation. All parameters to describe the influence of the time series sampling in Fourier space is entirely contained in the three parameters α_0, β_0, and θ_0. The values of α_0 and β_0 are measures for the maximum and minimum sig for all phase angles at a given frequency, and the angle θ_0 determines the phase angle where maximum sig is obtained at the frequency under consideration. A detailed description is given by Reegen (2007). If the keyword `profile` is provided in the `.ini` file, the sampling profile for the given time series is written to the file `profile.dat`. The four columns refer to

1. frequency [inverse time units],
2. α_0,
3. β_0,
4. θ_0 [rad].

Example. *In the file* `output.ini`*, the keyword* `profile` *is given and forces* SigSpec *to generate an output file* `output/profile.dat` *representing the sampling profile displayed in Fig.* 12.

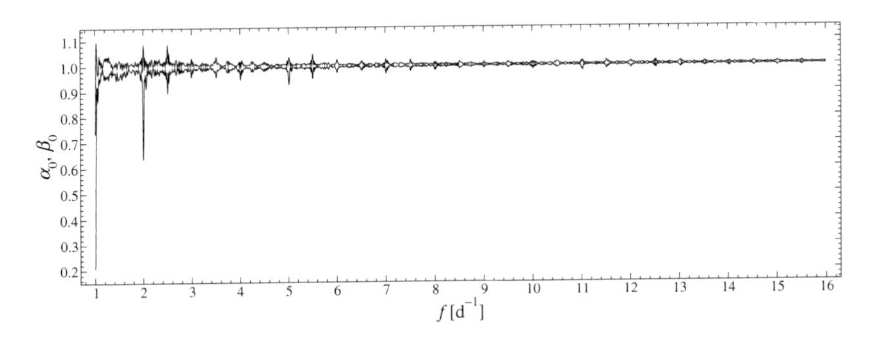

Figure 12: Sampling profile for the sample project `output`. The lower curve refers to α_0, the upper curve to β_0. The orientation angle of the rms error ellipse, θ_0 is not plotted.

5.3. Sock Diagram

The computation of a Sock Diagram is an optional add-on of SigSpec. If the keyword `sock:phases` is given in the `.ini` file, SigSpec computes *sock significances*, i. e. sig levels for a constant signal-to-noise ratio at a set of different phase angles, and for all frequencies for which spectra are calculated. As described by Reegen (2007), the expected sig level for a given amplitude signal-to-noise ratio at constant frequency and phase angle is proportional to the squared amplitude signal-to-noise ratio. Sig levels in the Sock Diagram are normalised to an expected value of 1, corresponding to an amplitude signal-to-noise ratio

$$\frac{A}{\langle A \rangle} = \frac{2}{\sqrt{\pi \log \mathrm{e}}} \approx 1.712 \,. \tag{10}$$

The sig level for an arbitrary signal-to-noise ratio may be deduced by multiplying the sig displayed in the Sock Diagram by $\frac{\pi \log \mathrm{e}}{4} \left(\frac{A}{\langle A \rangle}\right)^2 \approx 0.341 \left(\frac{A}{\langle A \rangle}\right)^2$.

Furthermore, the phase angle in the Sock Diagram is given with respect to θ_0, i. e. the phases with maximum sock significance are consistently aligned to zero phase for all frequencies.

The number of phase angles in the interval $[0, \pi[$ to be taken into account for each frequency of the spectrum has to be given as an argument to the keyword `sock:phases` in the `.ini` file. The sig levels in the phase intervals $[0, \pi[$ and $[\pi, 2\pi[$ are symmetric according to

$$\mathrm{sig}\,(A, \omega, \phi) = \mathrm{sig}\,(A, \omega, \phi + \pi)\ \forall \phi \,, \tag{11}$$

but both given in the output file `sock.dat` for completeness. The result represents a three-dimensional polar diagram of the sampling properties of the time series input file. To enhance the corresponding plot resolution, the number of phases specified with the keyword `sock:phases` in the `.ini` file is scaled by the maximum sig for each frequency. For sig maxima ≤ 1, the specified number is used, for sig maxima between 1 and 2, the number is doubled, and so on.

To enhance the quality of Sock Diagrams produced by SigSpec, the keyword `sock:fill` can be provided to specify a fill factor (as a floating-point number following the keyword). It is used for adaptive oversampling of frequencies according to the differences of maximum sigs for consecutive frequencies. The fill factor is the (rounded) number of additional frequencies per unit of sig difference. In other words, providing `sock:fill 10` guarantees that the resolution of the resulting Sock Diagram along the sig axis does not exceed 0.1, and an appropriate combination of the keywords `sock:phases` and `sock:fill` produces a Sock Diagram that mimics a continuous surface when plotted in 3D style. The default argument of `sock:fill` is 0, which means that adaptive oversampling is switched off.

Caution: the Sock Diagram may become a huge file! Especially for very low frequencies, a tremendous amount of data may be expected. Thus it is advisable either to exclude this frequency region (keyword `lfreq`) or to assign very low values to `sock:phases` and `sock:fill`.

The user may choose to obtain the Sock Diagram in three-dimensional cylindrical (default, or keyword `sock:cyl`) or cartesian coordinates (keyword `sock:cart`).

In any case, the output file `sock.dat` consists of three columns. In cylindrical coordinates, the columns refer to

1. height coordinate: frequency [inverse time units],
2. azimuthal coordinate: phase with respect to the sig maximum [rad],
3. radial coordinate: sock significance.

In cartesian coordinates, the columns refer to

1. frequency [inverse time units],
2. sock significance component in the direction of the sig maximum,
3. sock significance component in the direction of the sig minimum.

The keywords `sock:colmodel:lin` and `sock:colmodel:rank` permit to choose between two different colour models assigning RGB colours to the data points of the Sock Diagram. The linear model (`sock:colmodel:lin`) uses the sock significance as it is for colour scaling, whereas the rank model (`sock:colmodel:rank`) relies on a rank statistics of sock significances.

Caution: the computation of ranks may be very time-consuming! The progress control displayed during the calcucation of the rank statistics does not provide linear percentages in time. The percentage values refer to the portion of ranks among the number of data points that are finished.

A sequence of keywords `sock:colour` determines a colour path that is assigned to the data points in the Sock Diagram. The keyword is followed by four floating-point arguments. The first three arguments refer to the three RGB channels. Using the linear model (`sock:colmodel:lin`), the fourth argument is the sock significance to which the given colour has to be assigned. For the rank model (`sock:colmodel:rank`), the fourth argument is a floating-point value in the interval $[0, 1]$ and determines the fractile of data points to which the given colour has to be assigned. A value of, e. g., 0.5 assigns the specified colour to the median of sock significances. SigSpec performs a linear interpolation along this colour path and assigns a fourth column to the output file `sock.dat` containing RGB values. For entries beyond the start or end of the colour path, the start or end colour is used, correspondingly.

Example. *A linear colour model that produces colours from white via red, yellow, green, cyan, blue, and magenta to black is produced by the following specifications:*

```
sock:colmodel:lin
sock:colour 255 255 255 .5
sock:colour 255   0   0 .9
sock:colour 255 255   0 .95
sock:colour   0 255   0 1
sock:colour   0 255 255 1.05
sock:colour   0   0 255 1.1
sock:colour 255   0 255 1.2
sock:colour   0   0   0 2
```

Example. *A rank colour model producing greyscale coding is obtained by:*

```
sock:colmodel:rank
sock:colour   0   0   0 0
sock:colour 255 255 255 1
```

Example. *The Sock Diagram in the sample project* `output` *is generated according to the following entries in the file* `output.ini`*:*

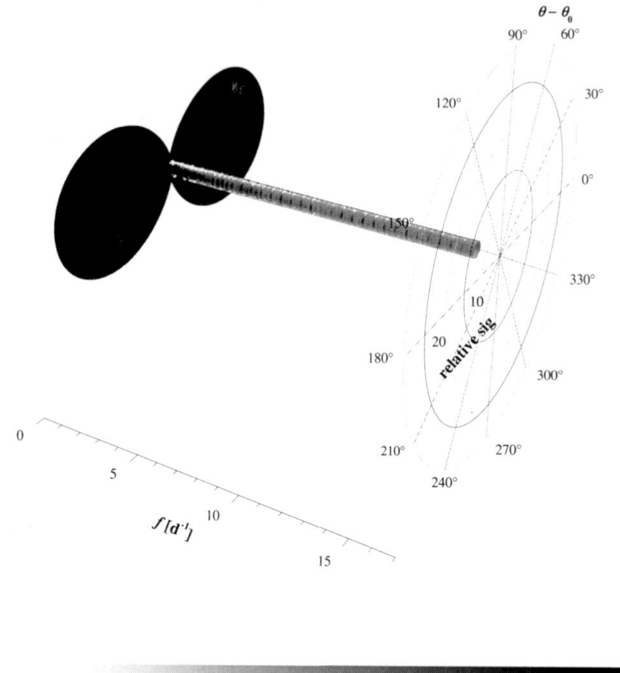

Figure 13: Sock Diagram for the sample project output.

```
sock:cyl
sock:phases 45
sock:fill 10
sock:colmodel:lin
sock:colour 255 255 255 0.98
sock:colour   0   0   0 1.02
```

The resulting file `output/sock.dat` *is displayed in Fig.* 13.

5.4. Phase Distribution Diagram

In addition to the spectral window and Sock Diagram, SigSpec can compute the probability density of phase angles at given frequency as a function of frequency. This is an alternative way to examine the properties of the sampling in the time domain and activated by the keyword `phdist:phases` in the `.ini` file. The resulting probability densities are normalised in a way that their mean over all phase angles is $\frac{1}{2\pi}$.

The number of phases to be computed is increased according to the eccentricity of the phase distribution at a given frequency.

In perfect analogy to the Sock Diagram (p. 31), there are further keywords available to adjust the contents of the output file `phdist.dat`.

- `phdist:fill` determines a filling factor for additional frequencies if the changes between the phase distributions for two adjacent frequencies are too high.
- `phdist:cyl` specifies cylindrical coordinates (height: frequency, azimuth: phase, radial: probability density of phase)
- `phdist:cart` specifies cartesian coordinates
- `phdist:colmodel:lin`
- `phdist:colmodel:rank`
- `phdist:colour`

Please refer to "Sock Diagram" (p. 31) for a detailed description.

Caution: For frequencies close to zero, tremendous output may be expected! Try to avoid the very low frequency region, if possible.

Example. *The Phase Distribution Diagram in the sample project* `output` *is generated according to the following entries in the file* `output.ini`*:*

```
phdist:cart
phdist:phases 30
phdist:fill 50
phdist:colmodel:rank
phdist:colour 223 223 223 0
phdist:colour  31  31  31 1
```

The resulting file `output/phdist.dat` is displayed in Fig. 14.

6. MultiSine Output

After each step of prewhitening, SigSpec performs a MultiSine least-squares fit over all significant signal components detected so far. Two optional types of output may help the user comprehend how this procedure performs at runtime.

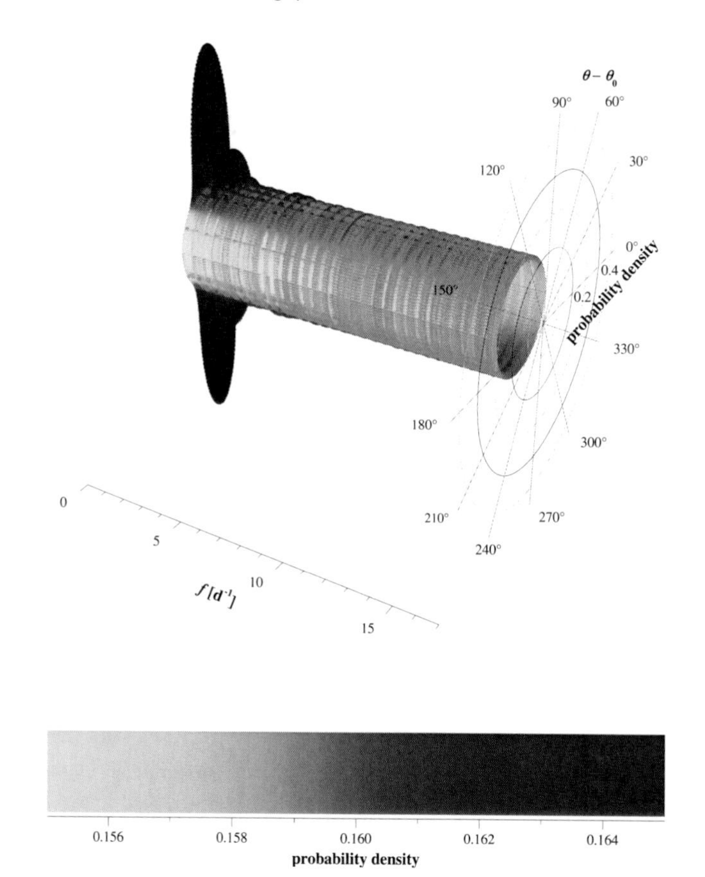

Figure 14: Phase Distribution Diagram for the sample project `output`.

6.1. MultiSine tracks

The MultiSine tracks allow to examine the changes in frequency, amplitude and phase of each signal component in the prewhitening cascade and are an alternative representation of the result files. Instead of a file index that refers to the iteration, the file index of the MultiSine track files `m#index#.dat` refers to the index of the component in the result files and lists its

1. frequency [inverse time units],
2. amplitude [units of observable],
3. phase [rad]

for each prewhitening step. In other words, a result file displays all the components for an iteration, whereas the MultiSine track file displays all the iterations for a component. Thus the MultiSine track provides a good estimate for the reliability and accuracy of the components found significant.

If the keyword `mstracks` is provided in the `.ini` file, MultiSine track files `m#index#.dat` are generated, where the index `#index#` starts with 000001, denoting the first significant signal component.

The keyword `mstracks` expects two integer parameters. The first defines the number of iterations for which entries in the MultiSine track files shall be generated. A negative number causes SigSpec to generate entries for all iterations. The second parameter has to be a positive number and defines a step width. If it is set 1, a line in the MultiSine track files is generated for each iteration, if it is set 2, for every second iteration (starting with `r000002.dat`), and so on.

Example. *The sample project* `output` *uses the keyword* `mstracks` *in the file* `output.ini`, *namely*

```
mstracks -1 1
```

providing MultiSine tracks `m000001.dat`, `m000002.dat`,..., *for all iterations of the entire prewhitening sequence. The MultiSine track for the primary signal component (file* `m000001.dat`*) is displayed in Fig.* 15.

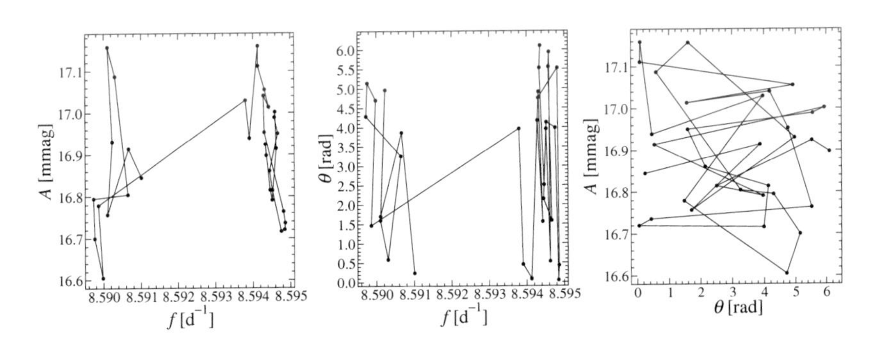

Figure 15: MultiSine track of the most dominant signal component in the light curve of the sample project `output` (8.59 cycles per day), according to the output file `output/m000001.dat`. *Left*: amplitude vs. frequency. *Mid:* phase vs. frequency. *Right:* amplitude vs. phase.

6.2. MultiSine profiles

A closer examination of the accuracy of the MultiSine fitting procedure is provided by the MultiSine profiles. If the user specifies the keyword `msprofs` in the `.ini` file, SIGSPEC produces additional output files `f#iteration#.dat`, `a#iteration#.dat`, and `p#iteration#.dat`. The idea is to evaluate the rms residual through modifying a single parameter of a single signal component, keeping all other parameters constant. Performing this operation for the frequency of each component produces a set of rms-residual-vs.-frequency plots, all written to the file `f#iteration#.dat`. Correspondingly, the files `a#iteration#.dat` and `p#iteration#.dat` contain rms-residual-vs.-amplitude and rms-residual-vs.-phase plots. The frequencies are scanned around the best fit by $\pm\frac{1}{T\sqrt{\mathrm{sig}}}$, T denoting the time interval width of the input time series, and sig referring to the signal component under consideration. The amplitudes are calculated from zero to twice the amplitude of best fit, and the phases in a range of $\pm\pi$ around the phase of best fit.

The keyword `msprofs` is followed by three integer values, the first denoting the number of data points an individual MultiSine profile shall consist of.[5] The second parameter defines the number of iterations for which MultiSine profiles shall be generated. A negative number causes SIGSPEC to generate profiles for all iterations. The third parameter has to be a positive number and defines a step width. If it is set `1`, a MultiSine profile is generated after each iteration, if it is set 2, after every second iteration (starting with `f000002.dat`, `a000002.dat`, `p000002.dat`), and so on.

The output files consist of seven columns:

1. frequency, amplitude, or phase, respectively,
2. rms residual,
3. first-order approximation, based on the tangential gradient (which should be zero, so that the deviation from zero is a measure of the accuracy of the MultiSine fitting procedure),
4. second-order approximation, based on the first and second derivatives at the parameter value of best fit,
5. point-to-point scatter,
6. index of the signal component,

[5] Due to the internal accuracy of the index computation, the actual number of points may differ from this value by ± 1.

7. index of the harmonic (0 for fundamental), see "Analysis of Harmonics", p. 51.

For each signal component, the first row refers to the parameter value of best fit, as used in the result file.

Example. *The sample project* output *uses the keyword* msprofs *in the file* output.ini, *namely*

```
msprofs 10000 50 3
```

*providing MultiSine profiles (*f000003.dat, a000003.dat, p000003.dat*), (*f000006.dat, a000006.dat, p000006.dat*),..., for a maximum of 50 iterations of prewhitening sequence. Each MultiSine profile is specified to contain 10 000 data points. The number of significant components found in the time series* output.dat *is 33, so that the last set of MultiSine profiles (*f000033.dat, a000033.dat, p000033.dat*) refers to the final solution contained in the file* result.dat*. The MultiSine ptofiles for the primary signal component at 8.59 cycles per day are displayed in Fig. 16.*

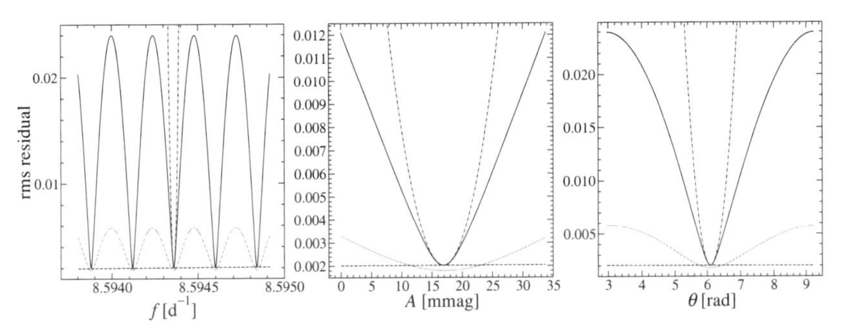

Figure 16: MultiSine profiles of the most dominant signal component in the light curve of the sample project output (8.59 cycles per day), according to the output files output/f000033.dat, output/a000033.dat, output/p000033.dat. *Left*: rms residual vs. frequency. *Mid:* rms residual vs. amplitude. *Right:* rms residual vs. phase. The *solid black* line refers to the rms residual, the *dashed black* line to the tangential gradient at the value of best fit (which should be zero), and the *dashed-dotted black* line to a second-order approximation based on the first two derivatives of rms residual. The *solid grey* line represents the point-to-point scatter.

7. Preview

Since the prewhitening cascade performed by SigSpec may be extremely time consuming, the program can compute a preview. This add-on is activated by the keyword preview in the .ini file.

Whereas the significance spectra rely on the False-Alarm Probability compared to a noise dataset with the same rms error as the given time series (or series of residuals, respectively), the significance spectrum provided in the file preview.dat represents a set of identified maxima in the significance spectrum of the original time series, but based on the point-to-point scatter in the time domain rather than on the standard deviation of observables. The lower sig limit for writing a local maximum to the file preview.dat is specified as the argument to the keyword preview in the .ini file.

The calculation of the sig is based on the assumption that only the point-to-point scatter is random, and everything else contributing to the rms error represents signal that will be prewhitened in the course of the subsequent loop. The preview output is to be considered as a rough estimate for the final result obtained by step-by-step prewhitening and contains not only the intrinsic variations but also all aliases, which will not occur in the following analysis. The file preview.dat consists of four columns referring to

1. frequency [inverse time units],
2. sig,
3. DFT amplitude [units of observable],
4. phase [rad].

Example. *The sample project* preview *contains a preview file for the V photometry of IC 4996 # 89. In the file* preview.ini, *the line*

```
preview:siglimit 5
```

sets the sig threshold to 5. The output file preview/preview.dat *contains 11 components, sorted by frequency. The frequencies and corresponding sigs in the first two columns are*

```
 0.9945158494303480    6.1674140356166323
 1.9917563998155081    6.9302735632389876
 2.1388902515119841    6.7175642729893710
 2.9835475482878717    8.6773027802854656
 3.1361308018982843    9.4899187898938777
 3.9862375005883859    8.9589776551282210
 4.1333713522847031    8.7492615592402885
 4.9780286490607102    5.0523760377159039
 5.1360613045861099    5.3438911274207790
11.0268647743572874    5.5214237212500406
12.0241053247411784    5.6674302270769710
```

Fig. 17 displays the significance and amplitude spectrum of the original time series. Since the preview does not employ any prewhitenings, aliases are present in the file.

- *The signal at 3.132 cycles per day corresponds to components # 3, 5, 7, and 9.*
- *The signal at 3.986 cycles per day corresponds to components # 1, 2, 4, 6, and 8.*
- *The signal at 5.409 cycles per day is not found in the preview. In the result of the prewhitening sequence, its sig is 5.02. Since the sig in the preview relies on the rms deviation of the original time series, whereas the final sig is based on the rms deviation after the previous prewhitening step, the sig associated to this frequency falls below the pre-selected threshold of 5 in the preview. The significance spectrum (grey line in the left panel of Fig. 17) shows a peak at the frequency under consideration the sig of which is $\approx$ 4.8.*
- *Components # 10 and 11 are 1-cycle-per-day aliases of each other, but do not show up in the final result,* `preview/result.dat`.

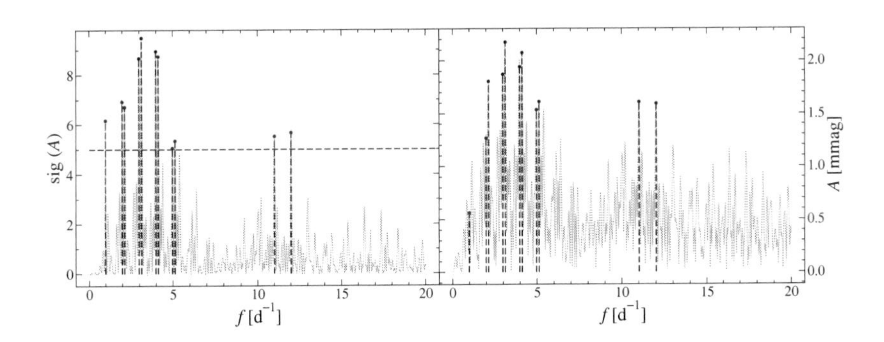

Figure 17: *Grey:* Fourier spectra for the sample project `preview`. *Left:* significance spectrum. *Right:* DFT amplitudes. The significant components in the preview are indicated by *dots* with dashed drop lines (file `preview/preview.dat`). The default sig threshold of 5 is represented by a horizontal *dashed line* in the left panel.

8. Correlograms

SigSpec is able to compute correlograms of the time series for each stage of prewhitening. The correlogram files are named `c#iteration#.dat`. The calculation of correlograms is activated by the keyword `correlograms`, which requires three integer parameters. The first parameter represents the maximum order to which to compute serial correlations, i. e. the limit of index lag for each correlogram. Setting it zero forces SigSpec to adjust it to half the number of data points in the time series. The second parameter is the maximum number of iterations for which to compute correlograms. If the number of prewhitening iterations exceeds this value, then no correlogram is generated for the iterations after this limit. If a number ≤ 0 is given, then a correlogram is computed for each prewhitening stage. The third parameter has to be a positive number and defines a step width. If it is set 1, a file is generated after each iteration, if it is set 2, after every second iteration (starting with `c000002.dat`), and so on. The correlogram computation is switched off by default.

A file `rescorr.dat` is generated, if the keyword `correlograms` is specified, no matter which parameter constellation is chosen.

A correlogram file consists of two columns referring to

1. index lag,
2. serial correlation coefficient.

Example. *The sample project* `correlograms` *illustrates how correlograms are generated with* SigSpec *using the V photometry of IC 4996 # 89 as time series input file* `correlograms.dat`. *The file* `correlograms.ini` *contains the line*

```
correlograms 100 -1 1
```

which forces SigSpec *to evaluate correlograms with a maximum index lag of 100 (first parameter) for all iterations (negative value of second parameter). After each iteration, a correlogram is generated (third parameter). The output files*

```
correlograms/c000000.dat
correlograms/c000001.dat
correlograms/c000002.dat
correlograms/rescorr.dat
```

are generated as displayed in Fig. 18.

9. Time-resolved Analysis

In time-resolved mode, SigSpec performs an analysis for a set of time intervals rather than for the entire time series. An interval of width given by the keyword `timeres:range` is moved in steps the width of which is given by the keyword `timeres:step` from the beginning of the time series to the end.[6] Consecutive

[6] In general, the step width is slightly modified by the software to achieve time-resolved analysis over the entire time series.

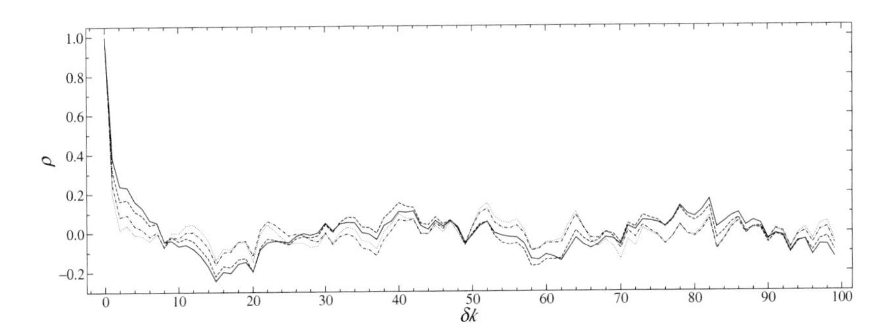

Figure 18: Correlograms for the sample project `correlograms`. *Solid:* correlogram of the initial time series (file `correlograms/c000000.dat`). *Dashed:* correlogram after one prewhitening (file `correlograms/c000001.dat`). *Dashed-dotted:* correlogram after two prewhitenings (file `correlograms/c000002.dat`). *Dotted:* residual correlogram after three prewhitenings (file `correlograms/rescorr.dat`).

time intervals are free to overlap. Time series data within such an interval are used to form a subset for which the analysis is performed. In addition, statistical weights may be applied to the subset data, all with respect to the centre of the interval, which shall be denoted t_C.

The only exception is the keyword `timeres:w:damp`. In this case, the analysis is optimised for signal excited at the beginning of the time interval corresponding to the subset under consideration, t_B and exponentially damped towards the end of the subset.

The weight functions of time are given in Table 1. The normalisation of weights is automatically performed by SIGSPEC. Also the combination of a weight function for time-resolved mode with weights columns (keyword `col:weights`) is supported.

In time-resolved mode, the set of output files as given in "Default Output", p. 24, is generated for each subset of the time series. This requires the introduction of an additional six-digit index, `#interval#`, in addition to `#iteration#`, and the annotation for the output files is

1. `wts.#interval#.dat` for the weight function vs. time in each subset,

2. `s#iteration#.#interval#.dat` for the spectra,

3. `t#iteration#.#interval#.dat` for the residuals after each step of prewhitening,

4. `r#iteration#.#interval#.dat` for the results after each step of prewhitening,

keyword	arguments	weight function
timeres:w:none		1
timeres:w:ipow	ξ	0 if $t = t_C$, $\lvert t - t_C \rvert^{-\xi}$ else
timeres:w:gauss	σ	$e^{-\left(\frac{t-t_C}{\sigma}\right)^2}$
timeres:w:exp	ζ	$e^{-\frac{\lvert t-t_C \rvert}{\zeta}}$
timeres:w:damp	ζ	$e^{-\frac{t-t_B}{\zeta}}$
timeres:w:cos	ν, Φ	$\cos\left(2\pi\nu \lvert t - t_C \rvert - \Phi\right)$
timeres:w:cosp	ν, Φ, ξ	$\cos^{\xi}\left(2\pi\nu \lvert t - t_C \rvert - \Phi\right)$

Table 1: Weight functions for time-resolved SigSpec analysis. The beginning of the time interval associated with the referring subset is denoted t_B, whereas t_C symbolises the centre of the time interval.

5. m#index#.#interval#.dat for the results after each step of prewhitening,
6. result.#interval#.dat for the result files, each with a list of significant signal components,
7. residuals.#interval#.dat for the final residuals after the prewhitening of all significant signal components,
8. resspec.#interval#.dat for the residual spectrum after the prewhitening of all significant signal components,

The column syntax is strictly consistent with the time-unresolved versions (see "Default Output", p. 24). The additional files, wts.#interval#.dat, are in two-column format. The first column represents the time values in the corresponding subset, the second column contains the weight function without normalisation.

Furthermore, SigSpec generates a file t000000.#interval#.dat, which contains the part of the original time series which is actually used as input.

Special functions – as introduced in "Analysis of the Time-domain Sampling" (p. 29), "Preview" (p. 39), and "Correlograms" (p. 41) – are also supplied with the #interval# index, i. e.

1. win.#interval#.dat for the amplitude windows,
2. profile.#interval#.dat for the sampling profiles,
3. sock.#interval#.dat for the Sock Diagrams,

4. `phdist.#interval#.dat` for the phase distribution diagrams,

5. `preview.#interval#.dat` for the previews,

6. `c#iteration#.#interval#.dat` for the correlograms after each step of prewhitening,

7. `rescorr.#interval#.dat` for the final correlograms after the prewhitening of all significant signal components.

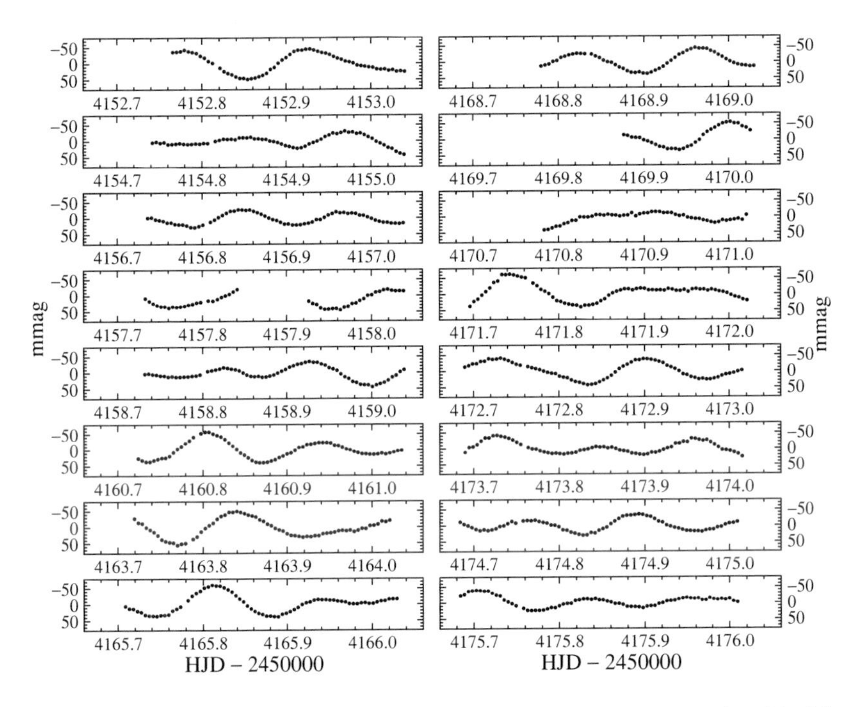

Figure 19: Time series used for the sample project `timeres`, representing 14 nights of Strømgren y photometry of the Delta Scuti star 4 CVn, acquired in February and March, 2007.

Example. *The sample project* `timeres` *illustrates the time-resolved analysis using Strømgren y photometry of the Delta Scuti star 4 CVn acquired with the Vienna University Automatic Photoelectric Telescope (Strassmeier et al. 1997). The data represent 16 nights from February 21 to March 16, 2007, and are displayed in Fig. 19.*

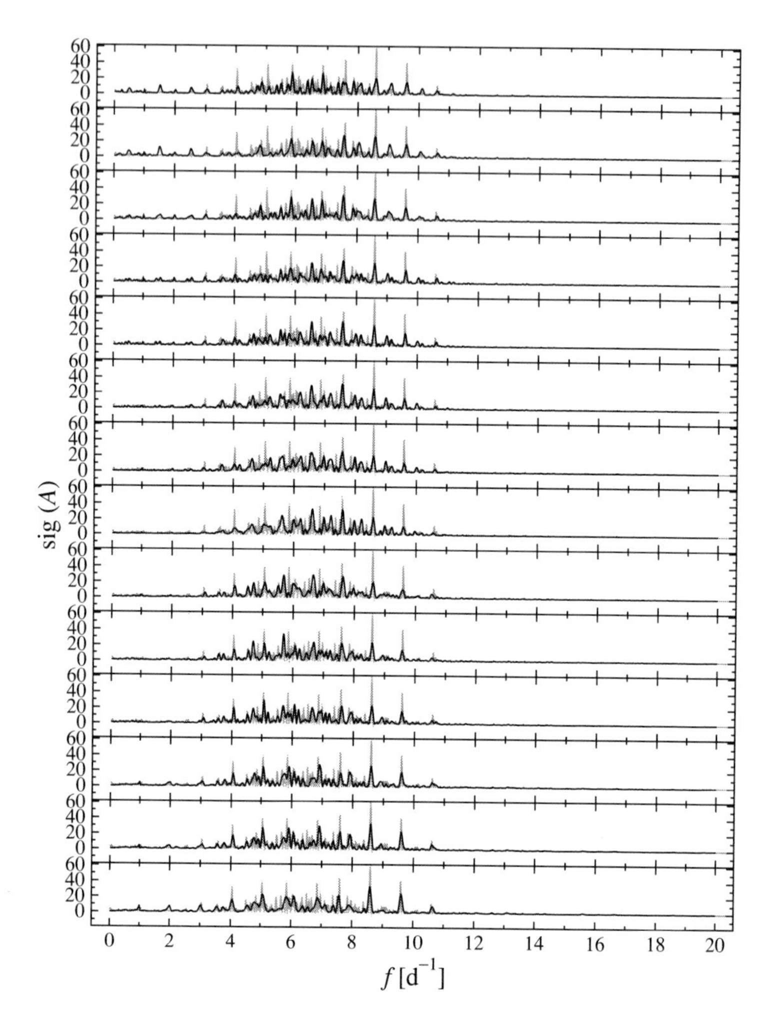

Figure 20: Time-resolved significance spectra for 14 subsets (from *top* to *bottom*) automatically generated in the sample project `timeres`. in each panel, the significance spectrum of the full dataset is displayed in *grey* colour for comparison.

The file `timeres.ini` *contains the specifications*

```
timeres:range 10
timeres:step 1
```

which provide a 10-day interval moving over the time base of 24 days, with one-day steps. The resulting 14 subsets are represented by the files `timeres/t000000.000000.dat` *to* `timeres/t000000.000013.dat`. *Gaussian weight functions with a standard deviation of 5 days are applied:*

```
timeres:w:gauss 5
```

The files `timeres/wts.000000.dat` *to* `timeres/wts.000013.dat` *contain the weights applied to each datapoint within each subset. Further output files are*

- `timeres/s000000.######.dat` *for the significance spectra of the original time series without prewhitening (Fig. 20) ,*
- `timeres/result.######.dat` *for the lists of significant signal components,*
- `timeres/residuals.######.dat` *for the residual time series after all prewhitening steps (divided into subsets according to the time intervals), and*
- `timeres/resspec.######.dat` *for the significance spectra of residuals.*

Here `######` *denotes six-digit numbers ranging from* 000000 *to* 000013.

10. SigSpec AntiAIC: Anti-aliasing Correction Mode

In AntiAIC mode, SigSpec does not follow a strict step-by-step prewhitening sequence. Instead, test runs are performed for a number of candidate peaks in the significance spectrum in order to find the solution that produces a minimum residual rms scatter after a user-given number of prewhitenings.

1. All peaks above a given sig limit are taken into consideration. The keyword `antialc:par` in the `.ini` file is followed by a floating-point number. This quantity is the AntiAIC parameter p_{al}, which has to attain a value in the interval $]0, 1]$. If the highest sig in the considered frequency range is $\max[\mathrm{sig}(A)]$, then the sig limit is $p_{\mathrm{al}} \max[\mathrm{sig}(A)]$. I. e., the AntiAIC parameter determines the sig limit for the candidate peak selection relative to the highest peak in the spectrum under consideration. Alternatively or in addition, a sig threshold for the AntiAIC candidate selection may be defined using the keyword `antialc:siglimit`. If neither `antialc:par` nor `antialc:siglimit` are present, the sig limit specified by `siglimit` in the `.ini` file (p. 23) is used for the AntiAIC candidate selection also.
2. The candidate selection is performed for each step in the test prewhitening sequence.

3. The resulting procedure is the computation of all combinations of candidate peaks above a sig threshold determined by the AntiAIC parameter. The number of iterations for these test prewhitenings is determined by the keyword `antialc:depth`, followed by an integer value. It specifies the depth of the AntiAIC computation.

4. The successful combination of peaks is selected upon the minimum residual rms deviation out of all examined combinations.

5. SIGSPEC does not necessarily adopt all iterations performed in the test run for the main prewhitening cascade. The integer value following the keyword `antialc:adopt` determines how many prewhitening steps shall be adopted. This quantity must not exceed the computation depth provided by the keyword `antialc:depth`. If the limits specified by the keywords `iterations`, `siglimit`, or `csiglimit` are reached, the output may even terminate before the number specified by the keyword `antialc:adopt`.

According to Reegen (2007), the expected sig is approximately proportional to the squared amplitude, if all influences by the time-domain sampling are neglected. The combination of n sinusoidal signal components interacting via aliasing is expected to produce a maximum amplitude that does not exceed the sum of amplitudes of the sinusoidal components. Consequently, the square root of the sig of such a combination, $\mathrm{sig}_{\mathrm{al}}$, is very likely below the sum of square roots of individual sigs sig_n,

$$\sqrt{\mathrm{sig}_{\mathrm{al}}} < \sum_n \sqrt{\mathrm{sig}_n} \, . \tag{12}$$

If these all are assumed equal and denoted $\mathrm{sig}_{\mathrm{ind}}$, then the upper sig limit for the alias is $\mathrm{sig}_{\mathrm{ind}}\sqrt{n}$. In other words, if a given peak with a sig $\mathrm{sig}_{\mathrm{al}}$ is an alias of a combination of n signal components with unique sigs $\mathrm{sig}_{\mathrm{ind}}$, then the individual significances are probably higher than $\frac{\mathrm{sig}_{\mathrm{al}}}{\sqrt{n}}$. In terms of the AntiAIC parameter, one obtains

$$n \approx \frac{1}{\sqrt{p_{\mathrm{al}}}} \tag{13}$$

for the approximate number of signal components that can be assigned alias-free for a given AntiAIC parameter p_{al}. Based on these considerations, SIGSPEC evaluates the AntiAIC computation depth using the AntiAIC parameter, if the keyword `antialc:depth` is not provided in the `.ini` file.

The AntiAIC mode produces additional screen output, if a combination of candidate peaks yields a lower residual scatter than the previous minimum, a two-line screen message is returned. The first line is a set of indices. In the example below, the AntiAIC parameter (keyword `antialc:par`) is set 0.5, and the AntiAIC computation depth (keyword `antialc:depth`) is 3. Correspondingly, the first line of output applies to the first of altogether ten candidate peaks in the first iteration, the first out of three in the second iteration, and the first out of seven in the third iteration. This peak constellation produces an rms deviation of residuals as displayed in the second line of output (in the example 0.00 405 851). After finishing the test cascade, the number of iterations specified by the keyword `antialc:adopt` (in the present example, this number is 2) is adopted for the main cascade. The screen output produced by the main cascade is the same as for a normal SigSpec prewhitening cascade without AntiAIC. The files containing spectra and residuals, respectively, are updated each time the residual rms deviation improves.

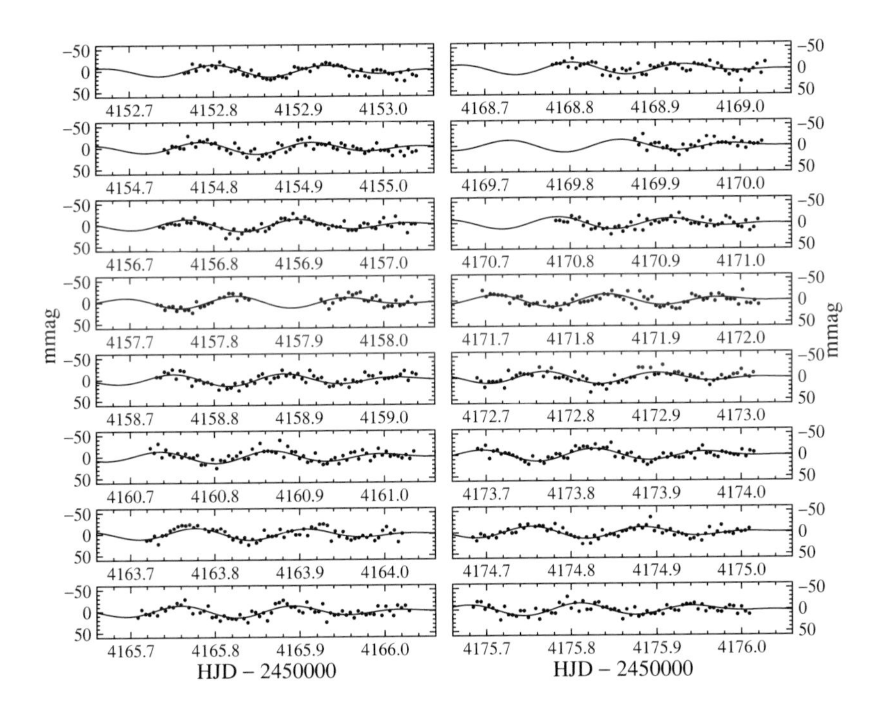

Figure 21: Time series used for the sample project `antialc` (*dots*). The sampling represents 14 nights of Strømgren y photometry of the δ Sct star 4 CVn, acquired in February and March, 2007. The magnitude values are synthesized forming two sinusoidal signals (*solid line*) plus Gaussian noise.

Example.[7] *The sample project* antialc *illustrates the anti-aliasing correction using the same sampling as the data for the sample project* timeres *(p. 45),*

1. *a sinusoid with frequency 6.5598 cycles per day, amplitude 7.29 mmag,*
2. *a sinusoid with frequency 8.5637 cycles per day, amplitude 6.87 mmag,*
3. *Gaussian noise with 7.36 mmag rms deviation,*

as displayed in Fig. 21 . The two signal frequencies differ by almost exactly 2 cycles per day and may easily be misidentified as aliases of each other. There are two identical versions of the light curve provided for comparison: alc.dat *and* antialc.dat.

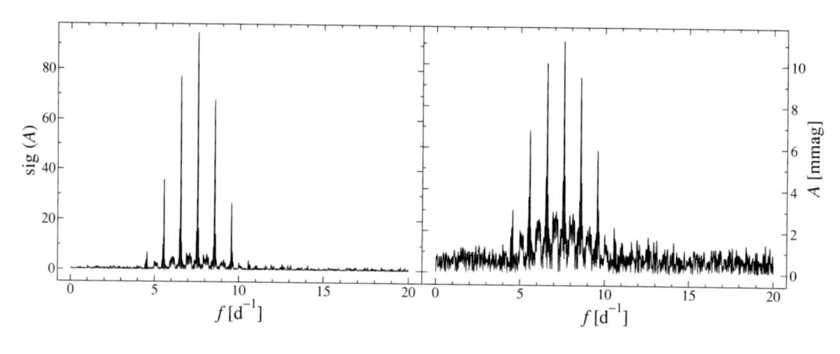

Figure 22: Fourier spectra for the sample project antialc. *Left:* significance spectrum. *Right:* DFT amplitudes.

The file alc.dat *corresponds to the project directory* alc, *representing a normal* SigSpec *run without a file* alc.ini. *Running* SigSpec alc, *the resulting frequencies (screen output) are*

```
1 freq 7.55917  sig 55.8792  rms 10.0617  csig 55.8792
2 freq 5.55706  sig 31.5539  rms 8.65888  csig 31.5539
3 freq 10.5668  sig 11.011  rms 7.81469  csig 11.011
4 freq 2.55231  sig 4.9934  rms 7.60001  csig 4.9934
```

Instead of the two signal components, 1-cycle-per-day aliases are identified. The significance and Fourier amplitude spectra of the dataset show the highest peak at 7.56 cycles per day, which represents a superposition of the first upper side peak of the signal at 6.56 cycles per day and the first lower side peak of the signal at 8.56 cycles per day (Fig. 22). This leads to an imperfect prewhitening of the two components, and the remaining signal is detected as a third component at 9.56 cycles per day.

[7] The computation of the sample project antialc takes 7 minutes on an Intel Core2 CPU T5500 (1.66GHz) under Linux 2.6.18.8-0.9-default i686.

The alternative AntiAIC analysis is provided by the file `antialc.ini`, *which contains the specifications*

```
antialc:par 0.5
antialc:depth 2
antialc:adopt 1
antialc:siglimit 4
```

All peaks that reach at least 50 % of the highest significance in the spectrum are taken into account. SigSpec *computes two consecutive iterations, but adopts only the first of these two iterations. A sig limit of 4 is assumed for the AntiAIC calculations (contrary to the default sig limit of 5 still valid as a breakup condition for the whole procedure). Running* `SigSpec antialc`, *the screen output is*

```
1 freq 6.55844  sig 55.0218  rms 10.0617  csig 55.0218
2 freq 8.56169  sig 43.6737  rms 8.68212  csig 43.6737
3 freq 33.7207  sig 3.97249  rms 7.48075  csig 3.97249
```

Both signals are recovered at a reasonable frequency accuracy. Moreover, according to the file `antialc/result.dat`, *the amplitudes of the two signals are recovered to a satisfactory precision (7.22 mmag, 6.47 mmag).*

11. Analysis of Harmonics

If a non-sinusoidal, but periodic process is measured, DFT does not only produce the fundamental frequency, which is the repetition rate of the non-sinusoid. The shape of the periodicity is recovered by a number of harmonics (also called overtones) the frequencies of which are integer multiples of the fundamental. In this case it may be considered insufficient to determine the exact frequency of the process by employing only the peak at the fundamental frequency and ignoring the harmonics. The keyword `harmonics`, followed by an integer determining the upper limit of the harmonic order, allows to compute the sig of the fundamental plus the desired number of overtones. The specification `harmonics 20` forces SigSpec to take into account altogether 21 frequencies.

As pointed out by Reegen (2007), SigSpec treats False-Alarm Probabilities in a statistically clean and unbiased way. In analogy to the comb analysis introduced by Kjeldsen et al. (1995), but benefitting from the exact statistical treatment of noise, it is possible to extend the method in order to evaluate the probability of a whole set of peaks to be generated by noise simultaneously. This strategy helps to take into account a fundamental frequency plus a set of integer multiples at once and permits to evaluate the most likely solution for a non-sinusoidal signal. In addition, the Fourier Space parameters obtained for the signal components provide a fit to the data in terms of a fundamental frequency plus overtones.

Given a set of amplitude levels A_h, $h = 0, 1, ..., H$, at different frequencies with associated False-Alarm Probabilities $\Phi_{\mathrm{FA}}(A_h)$, the probability that all amplitude levels are due to noise is given by the product of the individual False-Alarm Probabilities,

$$\Phi_{\mathrm{FA}}\left(\bigwedge_{h=0}^{H} A_h\right) = \prod_{h=0}^{H} \Phi_{\mathrm{FA}}(A_h) , \tag{14}$$

if the noise amplitudes at the two frequencies are assumed statistically independent. This is the probability that all amplitude levels are generated by noise.

Since the sig is defined as the negative logarithm of False-Alarm Probability, the above expression leads to

$$\mathrm{sig}\left(\bigwedge_{h=0}^{H} A_h\right) = \sum_{h=0}^{H} \mathrm{sig}(A_h) . \tag{15}$$

In this context, the sig represents the number of cases in one out of which all amplitude levels A_h are not generated by noise. This logical concept is the representation of an AND operator, as indicated by the argument to sig in the equation.

Reegen (2007) evaluated the expected value of the sig (ignoring the variations with frequency and phase) to be $\frac{\pi}{4} \log \mathrm{e} \approx 5.4575$. Considering H different amplitude levels simultaneously rescales this expected sig, so that we obtain $\frac{H\pi}{4} \log \mathrm{e}$. This rescaling may cause inconvenience, whence we use the *mean sig* of an individual peak out of this sample of fundamental plus harmonics,

$$\mathrm{msig}(A_h) := \frac{1}{H+1} \mathrm{sig}\left(\bigwedge_{h=0}^{H} A_h\right) , \tag{16}$$

instead. It is the expected sig obtained for an arbitrarily picked element out of the H peaks: if each of the considered peaks would have $\mathrm{msig}(A_h)$, then the total sig of the fundamental plus harmonics would be $\mathrm{sig}\left(\bigwedge_{h=0}^{H} A_h\right)$. The statistical properties of $\mathrm{msig}(A_h)$ are the same as for the "normal" sig analysis. If the keyword `harmonics` is provided in the `.ini` file, the sig levels returned in the second column of the file `result.dat` are mean sigs.

The result files display only the fundamentals of the solution, and information on the harmonics is stored in additional output files. The names are generated from the name of the corresponding result file without the extension `.dat`, plus `-h#index#.dat`, where `#index#` refers to the index of the item in the result file. For example, the harmonics for the third component in the file `result.dat` are stored in the file `result-h000003.dat`. The files contain the harmonics in ascending order, starting with the fundamental. The three columns are

1. sig of the individual peak,
2. DFT amplitude [units of observable],
3. Fourier-space phase angle [rad].

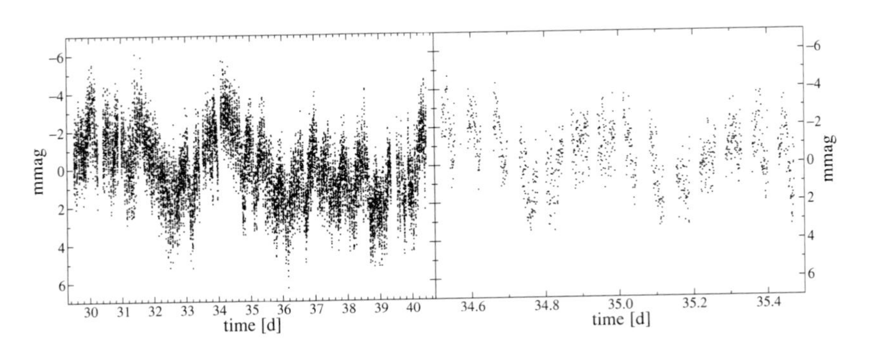

Figure 23: Time series used for the sample project harmonics.

Example. *The sample project* harmonics *illustrates the determination of a non-sinusoidal signal using the analysis of harmonics. The dataset represents (yet unpublished) space photometry of a star that exhibits surface activity. The task is to determine the rotation period of the star. For comparison, two identical versions of the time series are avalable (Fig. 23). The file* noharmonics.dat *is used together with the file* noharmonics.ini *to perform a* SigSpec *analysis without harmonics and associated with the project directory* noharmonics *containing the output. It contains four lines:*

```
ufreq 13
freqspacing .001
iterations 1
siglimit 0
```

In this constellation, SigSpec *computes the significance spectrum between 0 and 13 cycles per day, with steps of 0.001 cycles per day (Fig. 24, left panel). Only one iteration (i. e. no prewhitening) is performed. The highest peak is found at 0.296 cycles per day, which corresponds to a period of 3.38 days.*

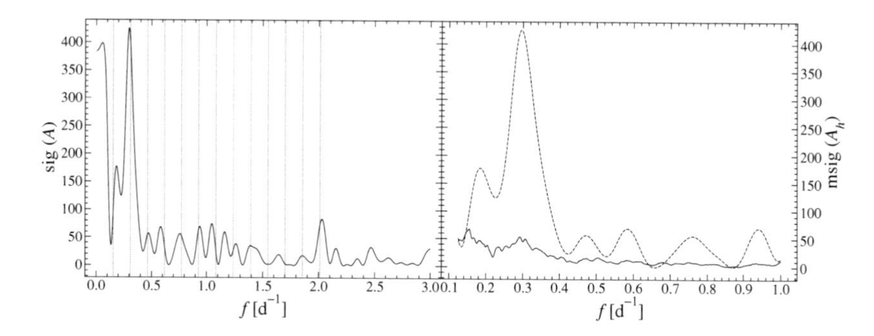

Figure 24: Fourier spectra for the sample project `harmonics`. *Left:* significance spectrum without employing the analysis of harmonics (*solid line*). The fundamental and twelve harmonics of the alternative solution are indicated by vertical *dashed lines* for comparison. *Right:* significance spectrum displaying the mean sig for fundamental plus twelve harmonics (*solid line*). Note that the frequency interval differs from the left panel. For comparison, the solution without harmonics is displayed as a *dashed line*.

The file `harmonics.dat` *is the same as* `noharmonics.dat`, *but the associated file* `harmonics.ini` *specifies a different setup by the lines*

```
lfreq 0.125
ufreq 1
freqspacing .001
iterations 1
siglimit 0
harmonics 12
```

It is advisable not to set the lower frequency limit zero, because below the Rayleigh frequency resolution, consecutive harmonics hit the same peak and produce unreliable results. In the present case, the Rayleigh frequency resolution is 0.091 cycles per day, and to be fairly on the safe side, the lower frequency limit is adjusted to 0.125 cycles per day. Fig. 24(right panel) contains the mean sig of the fundamental plus twelve harmonics vs. frequency.

The amplitudes of the fundamental and twelve harmonics are displayed vs. frequency in Fig. 25. The maximum sig is found at 0.155 cycles per day, i. e., the rotation period is 6.46 days, indicating that the analysis without harmonics led to a misidentification of the first harmonic as the "true" rotational frequency. For comparison, the left panel of Fig. 24 contains the fundamental plus harmonics found by this procedure as vertical dashed lines.

Moreover, for the analysis of harmonics, there is additional information in the screen output provided by SigSpec. *The standard screen output for the project* `noharmonics` *contains the lines*

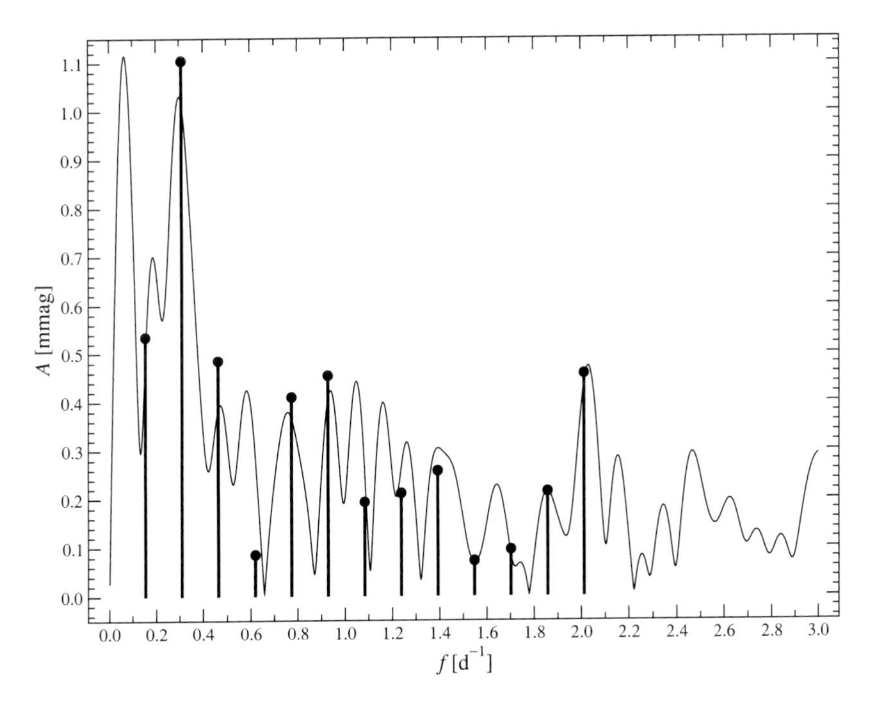

Figure 25: Frequencies and amplitudes of the harmonics associated to the most significant signal found for the sample project `harmonics` (*dots with drop lines*). The DFT amplitudes obtained by SigSpec without employing the analysis of harmonics are displayed as a *solid line* for comparison.

```
*** preparing to run SigSpec ******************************

Rayleigh frequency resolution              0.0914470160931467
oversampling ratio                        91.4470160931467433
frequency spacing                          0.0010000000000000
lower frequency limit                      0.0010000000000000
upper frequency limit                     13.0000000000000000
Nyquist coefficient                        0.9993990384615384
number of frequencies                  13000
```

For the project `harmonics`*, the corresponding output is richer.*

```
*** preparing to run SigSpec ******************************

Rayleigh frequency resolution              0.0914470160931467
oversampling ratio                        91.4470160931467433
frequency spacing                          0.0010000000000000
lower frequency limit                      0.1250000000000000
upper frequency limit                     13.0000000000000000
Nyquist coefficient                        1.0000000000000000
number of frequencies                  12876
```

```
upper fundamental frequency               1.0000000000000000
number of fundamental frequencies        876
```

Although the upper frequency limit is set 1 cycle per day by the keyword ufreq*,* SigSpec *has to compute the Fourier spectrum up to a frequency of 13 cycles per day in order to cover also the 12 harmonics. Two additional lines are provided corresponding to the upper limit for the fundamental frequencies, which is related to the specification by* ufreq *in the file* harmonics.ini*, and the number of fundamental frequencies.*

12. MultiFile Mode

12.1. How to handle multiple time series

An additional feature of SigSpec is the ability to handle multiple time series input files at once. This increases the performance of the program significantly, if the time values in all input files are identical.

- The user has to provide only one project directory <project> – just as in SingleFile mode (as described in "Projects", p. 7).
- Parameter specifications in the file <project>.ini are uniquely applied to all time series input files. Thus SigSpec expects the same column format for all time series input files and applies the settings specified in the .ini file to all input files.
- Time series files have to be indexed as #multifile#.<project>.dat, where #multifile# represents a six-digit index starting with 000000. Note that strictly ascending indices are required.
- All output files are supplied with the leading index #multifile#. For example, 000012.s000009.000002 denotes the significance specturm the second iteration for time interval number nine in a time-resolved analysis of the 12th file in MultiFile mode.
- The MultiFile mode is activated by the keyword multifile, followed by an integer value. This value is interpreted as the maximum index up to which the calculations shall be performed. This permits a restriction for, e. g., test runs. If the index limit is assigned a negative value, SigSpec analyses as many files as available.
- The sampling profile of the file 000000.<project>.dat is always written to a file. For subsequent and consistent time series, the sampling profile is taken from this file, which saves computation time. Only if the new

time values are inconsistent with those of the precursor, the profile is re-calculated and stored in a corresponding output file for later use. The keyword `profile` in the `.ini` file is ignored in MultiFile mode.

In MultiFile mode, SigSpec terminates, if a `#multifile#` index is reached, for which no time series input file is available.

A further keyword to restrict the MultiFile analysis is `mfstart`, which permits to specify a MultiFile index to start with (instead of 0).

Example. *The two lines*

```
mfstart 4
multifile 16
```

activate the MultiFile mode for input files from `000004.<project>.dat` *to* `000016.<project>.dat`.

The big advantage of the MultiFile mode is that sampling profiles are computed only if necessary. If the time-domain sampling is identical to a previously examined time series, the sampling profile of this time series is used. If the keyword `profile` *is set in the* `.ini` *file, a file* `assign.log` *is generated. It contains a table of assignments between time series file indices and profile indices.*

Example. *The line*

```
000013                              000002}
```

in the file `assign.log` *means that for* `000013.<project>.dat`*, the profile with index* `000002` *is used.*

Example. *The sample project* `multifile` *illustrates the simultaneous analysis of multiple time series. The project contains 10 time series files from* `000457.multifile.dat` *to* `000467.multifile.dat`*. However, the files do not represent a complete sequence, since* `000466.multifile.dat` *is missing. The lines*

```
mfstart 457
multifile 467
```

in the file `multifile.ini` *would force* SigSpec *to process the complete sequence of time series input files. Indeed, the program starts with the file* `000457.multifile.dat` *and proceeds until* `000465.multifile.dat`*. Since the next file,* `000466.multifile.dat` *is missing, it stops its calculations with* `000465.multifile.dat` *and displays a corresponding warning:*

```
Warning: MultiFile_Count 002
         MultiFile limit exceeds number of available
         time series input files, limit re-adjusted to 465.
```

The keyword `profile` *in the file* `multifile.ini` *forces* SigSpec *to generate the following files in the project directory:*

```
000457.profile.dat
000458.profile.dat
000460.profile.dat
000463.profile.dat
```

The reason why only four profiles are computed for nine time series is found in the file `assign.log`*:*

```
time series input file          profile and spectral window

        000457                          000457
        000458                          000458
        000459                          000457
        000460                          000460
        000461                          000457
        000462                          000457
        000463                          000463
        000464                          000457
        000465                          000457
```

The contents of this file tell the user that the samplings of all time series files are identical, except for those with indices 000458, 000460 *and* 000463. *In order to speed up the computations,* `SigSpec` *generates only one profile for the files with identical sampling and re-uses this profile for all of them. The first file with this sampling is* `000457.multifile.dat`*, and the associated profile is also used for*

```
000459.multifile.dat
000461.multifile.dat
000462.multifile.dat
000464.multifile.dat
000465.multifile.dat
```

If the keyword `win` *is added to the file* `multifile.ini`*, this assignment applies to the files containing the spectral windows as well.*

12.2. Differential significance spectra

Practical astronomical time series analysis occasionally comes along with target and comparison datasets that show coincident peaks in the DFT amplitude spectra. In this case, SigSpec provides a possibility to compute the probability that a peak in the target dataset is significant in spite of a given peak in the comparison dataset. Moreover, multiple target and/or comparison datasets may be handled the same way. The idea is to identify common (instrumental and/or environmental) effects and to distinguish them from periodicities exclusively found in a target dataset.

In the .ini file, there are three different keywords reserved for the specification of dataset types. Each expects one integer parameter representing the MultiFile index of the dataset under consideration.

1. The keyword `target` specifies a target dataset.
2. The keyword `comp` specifies a comparison dataset.
3. The keyword `skip` specifies a dataset to be ignored.

To enhance the convenience for the user, not all files need to be specified. The keyword `deftype` may be used to assign a default dataset type.

1. Use `deftype target` to assign the target attribute by default. If no `deftype` keyword is provided, this setting is activated.
2. Use `deftype comp` to assign the comp attribute by default.
3. Use `deftype skip` to assign the skip attribute by default.

Sampling profiles need to be computed for target datasets only. If the keyword `profile` is given in the .ini file, sampling profiles will only be generated for target datasets, and the file `assign.log` will also contain target datasets only.

To make datasets comparable even if their quality is different, the DFT spectra of the comparison datasets are scaled according to the power integral over the entire frequency range under consideration.

Instead of the observables c_k, $k = 0, 1, ..., K$ of a comparison dataset, the transformed quantities

$$c_k' := \frac{P(x_l)}{P(c_k)} c_k \tag{17}$$

are used, where x_l, $l = 0, 1, ..., L$ denotes the observables of the target dataset under consideration and P indicates the power integral of the quantity in parentheses. A DFT is calculated for each comparison dataset. There are two options to determine the resulting amplitude A_T to be compared to the target amplitude A.

By default, the sig measures the probability of a peak generated by noise at the same variance as that of the given time series. In case of computing differential sigs, the normalisation has to be modified, since part of the power found in the target spectrum is assumed due to corresponding power in a comparison spectrum. To take this into account appropriately, a factor

$$\gamma := \frac{P(x_l)}{dP} \tag{18}$$

is introduced, where dP is the power integral of the difference between the target data and the transformed comparison data. Correspondingly, the differential sig is a measure of the additional power with respect to the comparison dataset to be due to noise.

1. If the keyword `diff:comp` is set in the `.ini` file, a weighted arithmetic mean of the Fourier vectors, averaged over all comparison datasets is used to calculate A_T. The numbers of data points the comparison datasets consist of are used as weights. This option considers signal common among the comparison datasets only if the phases are aligned. Following the formalism by Reegen (2007), the cartesian representation of the differential sig evaluates to

$$\mathrm{sig}\left(a_{\mathrm{ZM}}, b_{\mathrm{ZM}} \mid \omega\right) = \gamma \frac{K \log \mathrm{e}}{\langle x^2 \rangle} \times$$
$$\left\{ \left[\frac{\left(a_{\mathrm{ZM}} - a_{T\,\mathrm{ZM}}\right) \cos\theta_0 + \left(b_{\mathrm{ZM}} - b_{T\,\mathrm{ZM}}\right) \sin\theta_0}{\alpha_0} \right]^2 \right.$$
$$\left. + \left[\frac{\left(a_{\mathrm{ZM}} - a_{T\,\mathrm{ZM}}\right) \sin\theta_0 - \left(b_{\mathrm{ZM}} - b_{T\,\mathrm{ZM}}\right) \cos\theta_0}{\beta_0} \right]^2 \right\} . \quad (19)$$

2. If the keyword `diff:compalign` is set in the `.ini` file, a weighted arithmetic mean of the DFT amplitudes, averaged over all comparison datasets is considered as A_T. The numbers of data points the comparison datasets consist of are used as weights. This option considers signal common among the comparison datasets also if they lag in phase. The differential sig is obtained through

$$\mathrm{sig}\left(A \mid A_T\right) = \gamma \frac{K \left(A - A_T\right)^2 \log \mathrm{e}}{4 \langle x^2 \rangle} \left[\frac{\cos^2\left(\theta - \theta_0\right)}{\alpha_0^2} + \frac{\sin^2\left(\theta - \theta_0\right)}{\beta_0^2} \right] , \quad (20)$$

following the annotation introduced by Reegen (2007).

The default setting is `diff:off`, which switches off the computation of differential sigs.

Additional output is provided in the spectra (see p. 26), where columns 6 and 7 contain the DFT amplitudes and phases of the transformed comparison dataset, respectively.

Example. *The sample project* `diffsig` *illustrates the analysis of target and comparison time series using differential significance spectra. There are nine time series input files available, indexed from* 000038 *through* 000046. *The file* `diffsig.ini` *contains the lines*

```
mfstart 38
multifile -1
```

which forces SigSpec *to start with the file* `000038.diffsig.dat` *and compute all available datasets. In this case,* SigSpec *takes into account all files from* 000038 *to* 000046*. The two lines*

```
deftype target
comp 38
```

in the file `diffsig.ini` *define the file* `000038.diffsig.dat` *as a comparison dataset and the rest as targets. Thus differential significance spectra are calculated for all time series from* 000039 *through* 000045*, with respect to* 000038 *as comparison data. The calculation of differential sigs is activated by the line*

```
diff:compalign
```

in the file `diffsig.ini`*, which produces differential sigs without respect to phase lags between comparison and target signals. The computations are made faster by the lines*

```
ufreq 7
siglimit 0
iterations 1
```

The sampling of the input file `000038.diffsig.dat` *represents the V photometry of IC 4996 # 89 (see Example* `SigSpecNative`*,* *p. 8**), and the observable is a synthetically generated signal with unit amplitude at a frequency of 3.125 cycles per day, plus Gaussian noise with 5 units rms deviation. The corresponding significance spectrum, as obtained by typing*

```
SigSpec 000038.diffsig
```

is displayed in the bottom panel of Fig. 26*. The five upper panels contain the differential significance spectra of the time series* 000039 *to* 000046*. These datasets contain 11 649 points and are based on the sampling used in the project* `harmonics` *(**p. 53**). Gaussian noise with a standard deviation of 100 units is generated. Just as in case of the comparison data, a sinusoid at 3.125 cycles per day is synthesized, but the phase is not the same as for* `000038.diffsig.dat`*. The amplitudes of this signal are 5 units for* 000039*, 6 units for* 000040*, 7 units for* 000041*, 8 units for* 000042*, 9 units for* 000043*, 10 units for* 000044*, 11 units for* 000045*, and 12 units for* 000046*. With increasing signal amplitude in the target data, the differential sig of the main peak consistently increases. In Fig.* 26 *the datasets* 000039 *to* 000046 *are displayed from bottom to top.*

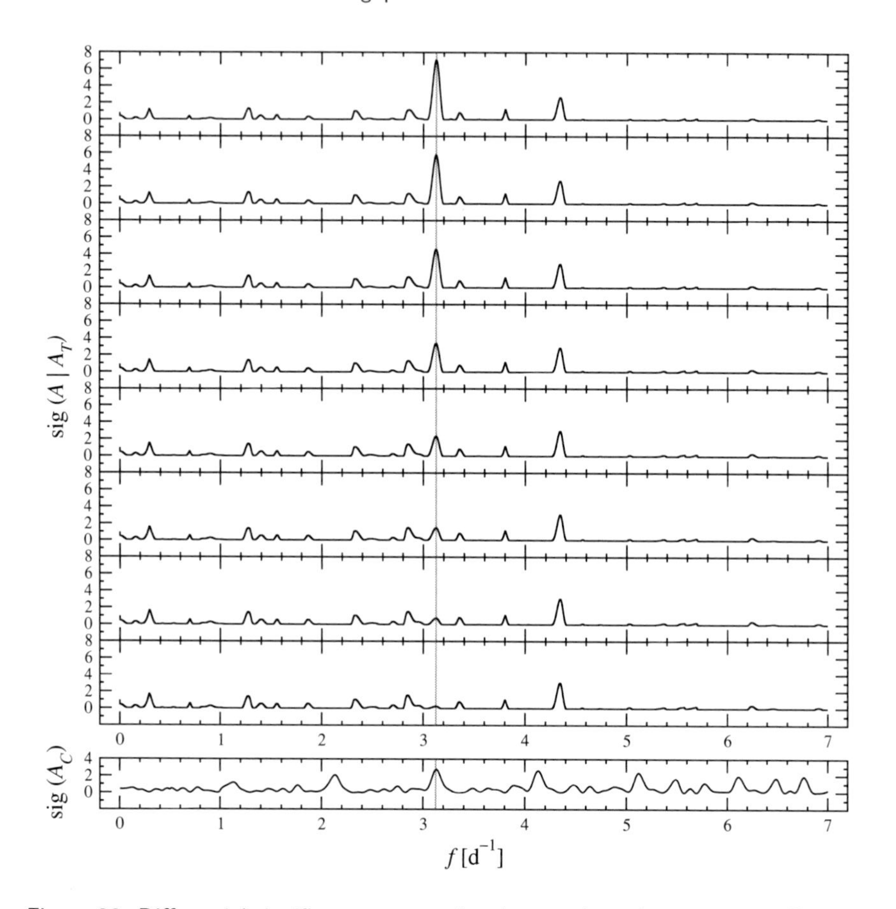

Figure 26: Differential significance spectra for the sample project `diffsig`. *Bottom:* significance spectrum of comparison data, representing a sinusoidal signal at 3.125 cycles per day (*grey line*), plus Gaussian noise. *Top eight panels:* Differential significance spectra for target time series representing the Gaussian noise plus a sinusoidal signal at 3.125 cycles per day. Both the time-domain sampling and the signal phase differ from the comparison data. From bottom to top, the amplitude of this signal increases.

13. The Built-in Simulator

SigSpec contains a simulator to generate and analyse synthetic time series. To activate the simulator, a sequence of keywords may be given in the `.ini` file to generate a variety of datasets. The sampling is taken from the time series input file.

The simulator activities specified by `sim:signal`, `sim:poly`, `sim:exp`, sim:serial, sim:temporal, sim:rndsteps, and sim:zeromean are interpreted as a sequence and performed step by step, following their order in the .ini file. SigSpec generates the synthetic light curve by performing all specified actions following the order of occurrence in the .ini file.

The synthetic time series is saved as a file with the same name as the input, but in the project directory, to avoid accidential overwriting of original data. If the time series input file is named <project>.dat, then the synthetic time series is <project>/<project>.dat. In MultiFile mode, if the time series input files are named #multifile#.<project>.dat, the synthetic time series are <project>/#multifile#.<project>.dat.

13.1. The simulator mode

SigSpec supports two different simulator modes.

1. The keyword sim:add runs the simulator in additive mode. The program keeps the original observable values and adds the synthetic values. For example, this function is useful to add synthetic noise to a given time series.

2. The keyword sim:replace forces the simulator to overwrite the original observable values with the synthetic values.

3. The keyword sim:off is used, if no simulator activity is desired. Since the simulator is deactivated by default, this keyword is redundant and only implemented for completeness.

13.2. Random numbers

The SigSpec simulator is capable of modelling three different types of random processes:

- serially correlated noise (keyword sim:serial, p. 70),

- temporally correlated noise (keyword sim:temporial, p. 72),

- random steps (keyword sim:rndsteps, p. 74.

The random number generator employed for these models may be initialised in two different ways.

1. The user may pass an integer value to the program. This value has to be written into a file <project>.rnd.

2. If the file <project>.rnd is not present, the simulator initialises the random number generator using the system time.

The last integer value in the sequence of random numbers is written to a file `<project>/<project>.rnd`. This allows to embed SigSpec into an external loop for numerical simulations. If the output file `<project>/<project>.rnd` is moved to `<project>.rnd` externally between consecutive SigSpec runs, the program may used iteratively without breaking the random number sequence.

Example. *The simulator is employed in the sample projects* `sim-serial`, `sim-temporal` *and* `sim-rndsteps`. *To initialise the random number generator, a file* `sim-serial.rnd`, `sim-temporal.rnd` *and* `sim-rndsteps.rnd`, *respectively, is used to make the output reproducible.*

Consequently, the user has three options to explore the these samples.

1. *If the samples are processed as they are,* SigSpec *reproduces the given output exactly.*

2. *If the* `.rnd` *file in the input directory is removed by the user,* SigSpec *produces a new set of random numbers. The random number generator is initialised employing the system time.*

3. *If the content of the* `.rnd` *file in the input directory is modified by the user,* SigSpec *produces a new set of random numbers. The random number generator is initialised employing the new number in the* `.rnd` *file.*

13.3. Sinusoidal signal

The keyword `sim:signal` is given with five floating-point parameters. They specify

1. the lower time limit,

2. the upper time limit,

3. the amplitude,

4. the time zeropoint (a fixed time where the signal shall attain a maximum), and

5. the frequency [inverse time units].

If the lower and upper time limits are both set zero, the signal is generated for the entire time base.

Example. *The sample project* `sim-signal` *contains the simulation and analysis of two sinusoidal signals, one over the entire time base, one on a restricted time*

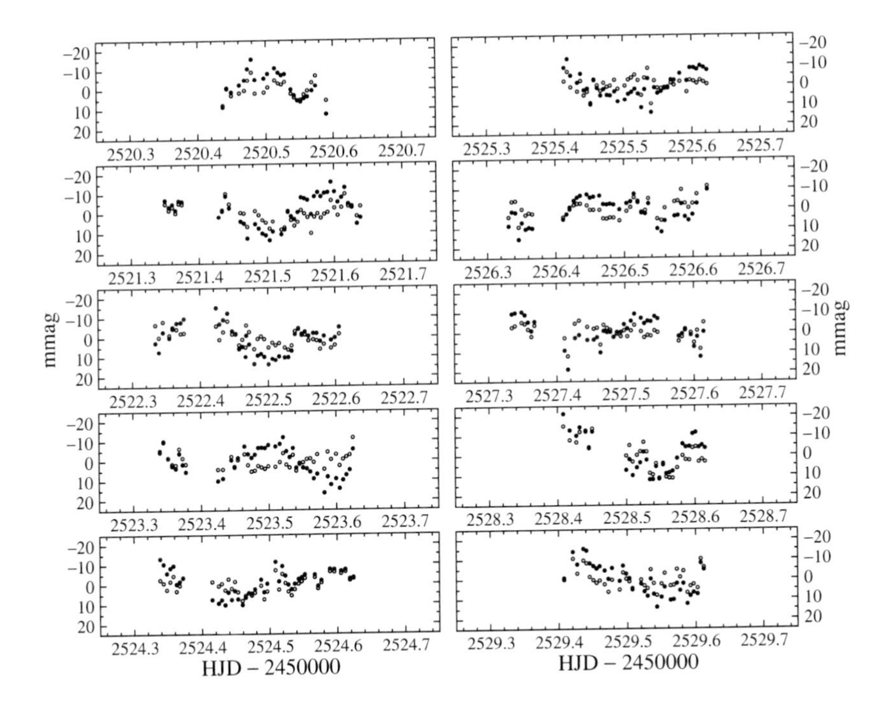

Figure 27: Time series generated by the simulator in the sample project `sim-signal`. *Open circles:* Original V photometry of IC 4996 # 89. *Dots:* Two sinusoidal signals added by the simulator.

interval. In this sample project, the V photometry of IC 4996 # 89 (see Example `SigSpecNative`*, p. 8) is modified, according to the line*

```
sim:add
```

in the file `sim-signal.ini`*. The line*

```
sim:signal 0 0 0.00727 2521.4542 4.68573
```

produces a sinusoidal signal over the entire time base (corresponding to the first two arguments being zero). The amplitude is 7.27 mmag, and the frequency is 4.68573 cycles per day. At HJD 2452521.4542 the sinusoid shall attain zero value. Correspondingly, the line

```
sim:signal 2521 2525 0.00543 2524.2356 6.24512
```

is associated to a sinusoid with amplitude 5.43 mmag, frequency 6.24512 cycles per day, and a zeropoint at HJD 2452524.2356. This signal is not generated for the entire time base but only from HJD 2452521 to HJD 2452525. Fig. 27 displays the light curves of the original and the synthetic data.

The screen output contains the lines

```
*** simulator: add ******************************************

signal
signal
```

indicating that the simulator adds the synthetic values to the original observables, and that two sinusoids are generated.

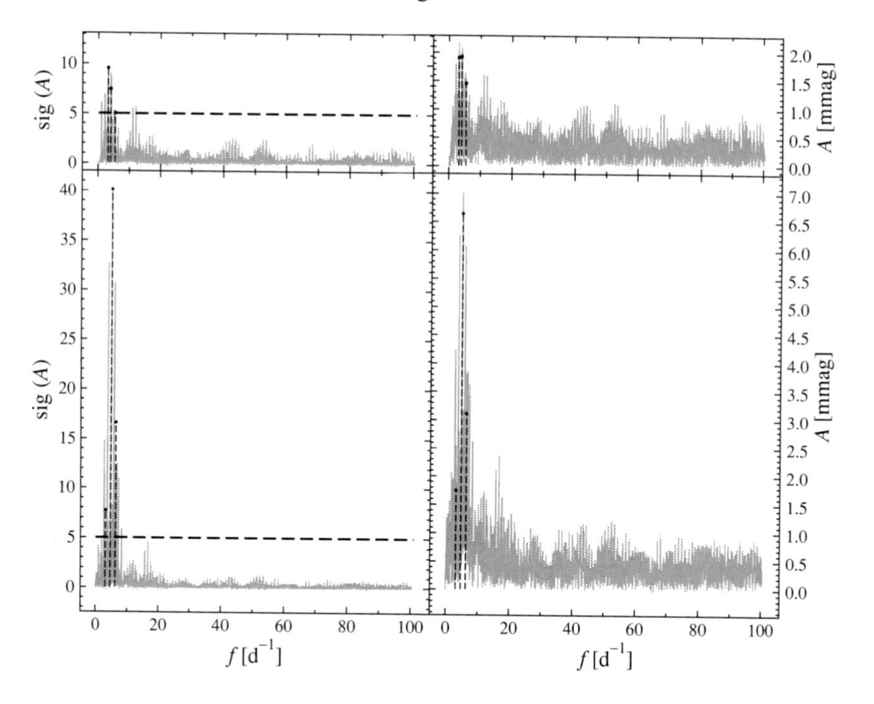

Figure 28: Fourier spectra for the sample project `sim-signal`. *Left:* significance spectra. *Right:* DFT amplitudes. *Top:* original spectra (file `SigSpecNative/s000000.dat`). *Bottom:* spectra with two sinusoidal signals added. All spectra are plotted *grey*. The significant components are indicated by black *dots* with dashed drop lines (file `SigSpecNative/result.dat` for the top panels, file `sim-signal/s000000.dat` for the bottom panels). The default sig threshold of 5 is represented by a horizontal *dashed line* in the left panels.

Fig. 28 compares the Fourier spectra of the synthetic time series to those of the original time series (as used in Example `SigSpecNative`, p. 8, and displayed in Fig. 2, *p. 11. Both signals introduced by the simulator are identified, but the prewhitening of the component at 6.25 cycles per day is performed over the whole time base, although the signal is present only in an interval. This introduces additional noise, which causes the signal at 3.99 cycles per day to drop below the significance limit of 5 and avoids the detection of the component at 5.41 cycles per day.*

13.4. Polynomial trend

The keyword `sim:poly` is given with five floating-point parameters. They specify

1. the lower time limit,
2. the upper time limit,
3. the coefficient P_0,
4. the time zeropoint t_0, and
5. the exponent X.

If the exponent is a non-integer number, the simulator evaluates

$$P(t) := P_0 \left|t - t_0\right|^X \tag{21}$$

instead and produces a power function.

For integer exponents, the trend is generated by the relation

$$P(t) := P_0 \left(t - t_0\right)^X \ . \tag{22}$$

Thus a full polynomial may be constructed by multiple keywords `sim:poly` with different parameters and integer exponents.

If the lower and upper time limits are both set zero, the polynomial trend is generated for the entire time base.

Example. *The sample project* `sim-poly` *contains the simulation and analysis of 16 individual power functions defined on different time intervals (Fig. 29, p. 68). The sampling of the V photometry of IC 4996 # 89 is used, and the simulator replaces the original observable values, according to the line*

```
sim:replace
```

in the file `sim-poly.ini`*. The specifications for the power functions are contained in the lines*

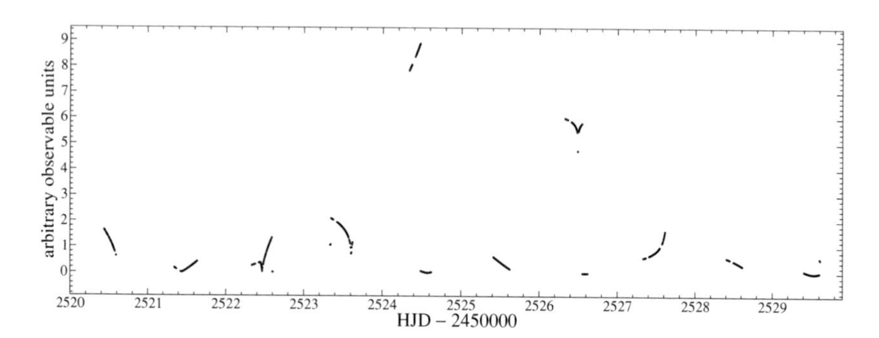

Figure 29: Time series generated by the simulator in the sample project `sim-poly`. The sampling represents the V photometry of IC 4996 # 89. The simulator replaces the origninal observable by 16 different power functions.

```
sim:poly   2520.215 2521.088 4.298 2520.626  0.581
sim:poly   2521.088 2521.679 2.932 2521.443  1.195
sim:poly   2521.679 2522.442 1.092 2522.067  1.063
sim:poly   2522.442 2522.595 5.372 2522.466  0.676
sim:poly   2522.595 2523.351 2.495 2522.682  2.042
sim:poly   2523.351 2523.924 2.839 2523.607  0.221
sim:poly   2523.924 2524.478 8.357 2525.412 -0.899
sim:poly   2524.478 2525.399 2.304 2524.576  1.432
sim:poly   2525.399 2526.107 2.573 2525.721  1.205
sim:poly   2526.107 2526.550 6.350 2526.493  0.031
sim:poly   2526.550 2526.847 4.192 2526.589  2.893
sim:poly   2526.847 2527.616 0.345 2527.652 -0.472
sim:poly   2527.616 2528.264 3.583 2527.783  0.725
sim:poly   2528.264 2528.777 1.246 2528.704  0.610
sim:poly   2528.777 2529.606 3.534 2529.535  1.752
sim:poly   2529.606 2530.242 9.002 2529.694  1.119
```

The screen output contains the lines

```
*** simulator: replace **************************************

polynomial trend
polynomial trend
polynomial trend
polynomial trend
polynomial trend
polynomial trend
polynomial trend
polynomial trend
polynomial trend
polynomial trend
polynomial trend
polynomial trend
polynomial trend
polynomial trend
polynomial trend
polynomial trend
```

to indicate that the simulator replaces the original observables by the synthetic values, and that 16 power functions are generated.

SigSpec *detects 19 significant signal components, which are not discussed here.*

13.5. Exponential trend

The keyword `sim:exp` is given with five floating-point parameters. They specify

1. the lower time limit,
2. the upper time limit,
3. the coefficient E_0,
4. the time zeropoint t_0, and
5. the exponent X.

The polynomial trend is generated by the relation

$$E(t) := E_0 \, \mathrm{e}^{X(t-t_0)} . \tag{23}$$

If the lower and upper time limits are both set zero, the exponential trend is generated for the entire time base.

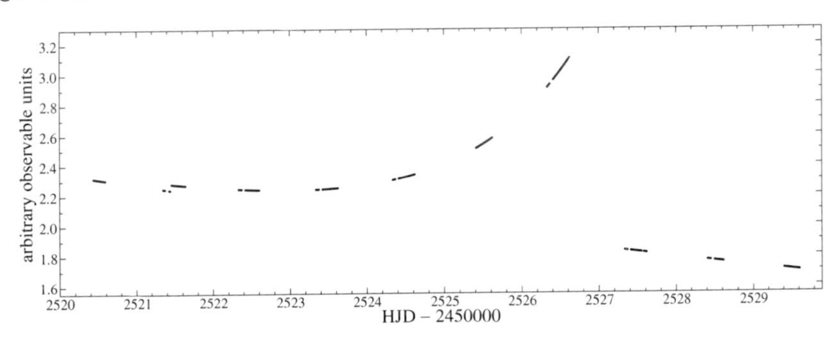

Figure 30: Time series generated by the simulator in the sample project `sim-exp`. The sampling represents the V photometry of IC 4996 # 89. The simulator replaces the origninal observable by two exponential functions, one over the entire time base, and the other one on an interval between HJD 2452521.4532 and HJD 2452526.8832.

Example. *The sample project* `sim-exp` *contains the simulation and analysis of two exponential trends, one over the entire time base, one on a restricted time interval, corresponding to the lines*

```
sim:exp 2521.4532 2526.8832 1.3256 2526.7384  0.65834
sim:exp    0          0       2.2841 2520.8562 -0.03425
```

in the file `sim-exp.ini`. *The sampling of the V photometry of IC 4996 # 89 is used, and the simulator replaces the original observable values, according to the line*

```
sim:replace
```

The screen output contains the expression `exponential trend` *to indicate that such a trend is generated. In this example, the entry is found twice. The resulting light curve is displayed in* *Fig. 30, p. 69.*

SigSpec *detects 54 significant signal components, which are not discussed here.*

13.6. Serially correlated noise

This simulator module produces Gaussian noise the standard deviation of which may vary in time according to a polynomial trend. A serial correlation coefficient between consecutive data points may be specified additionally.

The keyword `sim:serial` is given with six floating-point parameters. They specify

1. the lower time limit,
2. the upper time limit,
3. the coefficient σ_0 for the standard deviation of the Gaussian noise,
4. the time zeropoint t_0 for the polynomial trend of the standard deviation,
5. the exponent X for the polynomial trend of the standard deviation, and
6. the serial correlation coefficient.

The standard deviation of the Gaussian noise follows the relation

$$\sigma(t) := \sigma_0 \,(t - t_0)^X \; . \tag{24}$$

A full polynomial may be constructed by multiple keywords `sim:serial` with different parameters.

If the lower and upper time limits are both set zero, the noise is generated for the entire time base.

Example. *The sample project* `sim-serial` *contains the simulation and analysis of serially correlated noise. The sampling of the V photometry of IC 4996 # 89 is used, and the simulator replaces the original observable values, according to the line*

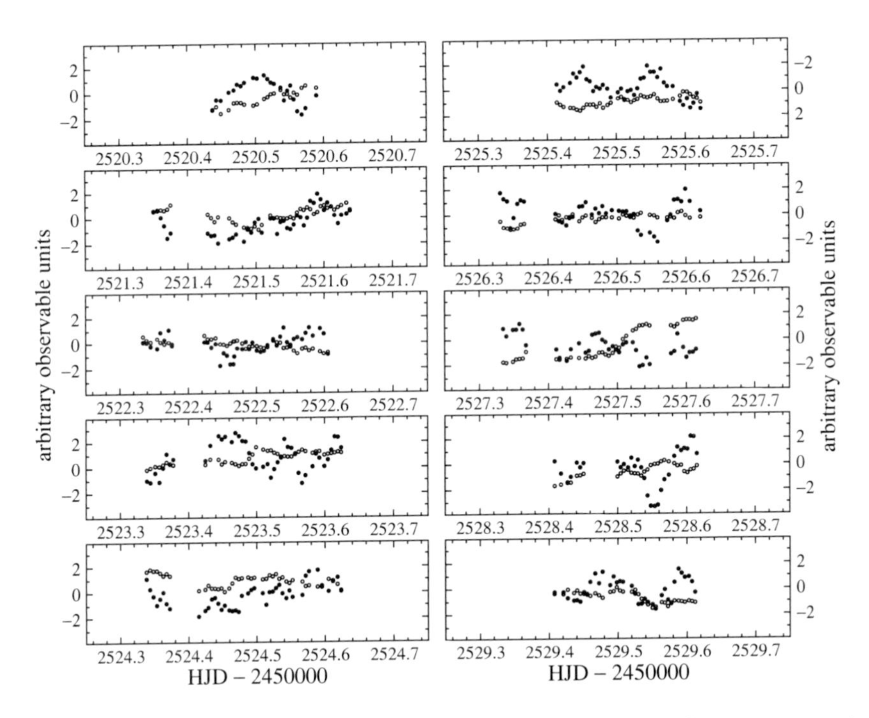

Figure 31: Time series generated by the simulator in the sample projects `sim-serial` (*dots*) and `sim-temporal` (*open circles*), respectively. The sampling represents the V photometry of IC 4996 # 89. In both samples, the original observable values are replaced by the simulator.

```
sim:replace
```

in the file `sim-serial.ini`. *The line*

```
sim:serial   0 0 1 0 0 0.8
```

specifies noise with a constant standard deviation of 1 and a serial correlation coefficient of 0.8. Setting the first two parameters zero provides synthetic data for the entire time series. The resulting light curve is displayed in Fig. 31. *The line*

```
random number generator: file sim-serial.rnd
```

in the screen output indicates that a file `sim-serial.rnd` *is found and used to initialise the random number generator. If such a file were not present, the system time would be used:*

```
random number generator: system time initialisation
```

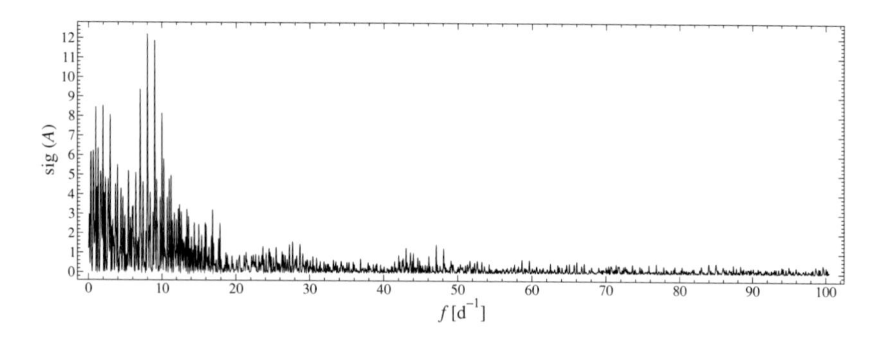

Figure 32: Typical significance spectrum for serially correlated noise, based on the sampling of the V photometry of IC 4996 # 89. Serial correlation produces systematically higher sigs in the low frequency region.

A significance spectrum is displayed in Fig. 32. The overall shape of the spectrum is typical for serially correlated noise, characterised by higher amplitudes and sigs for low frequencies.

13.7. Temporally correlated noise

This simulator module produces Gaussian noise the standard deviation of which may vary in time according to a polynomial trend. A temporal correlation coefficient R_T between consecutive data points t_{n-1}, t_n may be specified. In contrast to the serial correlation, the temporal correlation takes into account the width of the time interval between pairs of data points, which has implications on the noise behaviour of non-equidistantly sampled data. The serial correlation R_S drops exponentially with the distance in time according to

$$R_S := R_T^{t_n - t_{n-1}} . \qquad (25)$$

In this context, the temporal correlation coefficient may be interpreted as the serial correlation coefficient of two data points separated by one unit of time.

The keyword `sim:temporal` is given with six floating-point parameters. They specify

1. the lower time limit,
2. the upper time limit,
3. the coefficient σ_0 for the standard deviation of the Gaussian noise,
4. the time zeropoint t_0 for the polynomial trend of the standard deviation,

5. the exponent X for the polynomial trend of the standard deviation, and
6. the temporal correlation coefficient R_T.

The standard deviation of the Gaussian noise follows the relation

$$\sigma(t) := \sigma_0 \left(t - t_0\right)^X \ . \tag{26}$$

A full polynomial may be constructed by multiple keywords `sim:temporal` with different parameters.

If the lower and upper time limits are both set zero, the noise is generated for the entire time base.

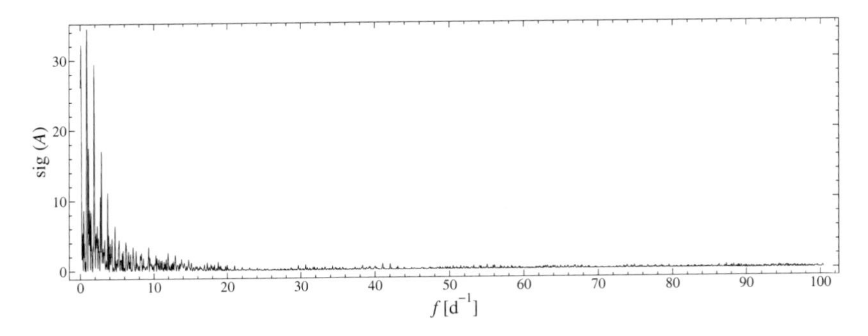

Figure 33: Typical significance spectrum for temporally correlated noise, based on the sampling of the V photometry of IC 4996 # 89. Temporal correlation produces systematically higher sigs in the low frequency region, which is quite comparable to serial correlation (Fig. 32).

Example. *The sample project* `sim-temporal` *contains the simulation and analysis of temporally correlated noise. The sampling of the V photometry of IC 4996 # 89 is used, and the simulator replaces the original observable values, according to the line*

```
sim:replace
```

in the file `sim-temporal.ini`. *The line*

```
sim:temporal   0 0 1 0 0 0.01
```

specifies noise with a constant standard deviation of 1 and a temporal correlation coefficient of 0.01. Setting the first two parameters zero provides synthetic data for the entire time series. The resulting light curve is displayed in Fig. 31. Comparing this light curve to the dataset generated in the project `sim-serial` *(p. 70), the correlation between consecutive data points is obviously much stronger in the present example. Using Eq. 25 with a typical sampling interval width of 9 min for the dataset under consideration, the temporal correlation coefficient of 0.01 corresponds to a serial correlation coefficient of* ≈ 0.97*.*

The significance spectrum displayed in Fig. 33 shows the same overall characteristics as the corresponding spectrum for serially correlated noise (Fig. 32, but the sigs at low frequencies are considerably higher, which is a consequence of the strong serial correlation associated to this setup.

13.8. Random steps

This module generates steps following two random processes:

1. the constant attained by the synthetic observable throughout each step follows a Gaussian distribution with an expected value 0,
2. a Poisson process is used to define when a step has to be incorporated.

The keyword `sim:rndsteps` is given with four floating-point parameters. They specify

1. the lower time limit,
2. the upper time limit,
3. the standard deviation of the Gaussian distribution defining the constants attained throughout each step,
4. the expected time range for the Poisson distribution of steps.

If the lower and upper time limits are both set zero, the steps are generated for the entire time base.

Example. *The sample project* `sim-rndsteps` *illustrates the simulation and analysis of random steps upon the sampling of the V photometry of IC 4996 # 89. The simulator replaces the original observable values, according to the line*

```
sim:replace
```

in the file `sim-rndsteps.ini`*. The line*

```
sim:rndsteps   0 0 0.5 0.07
```

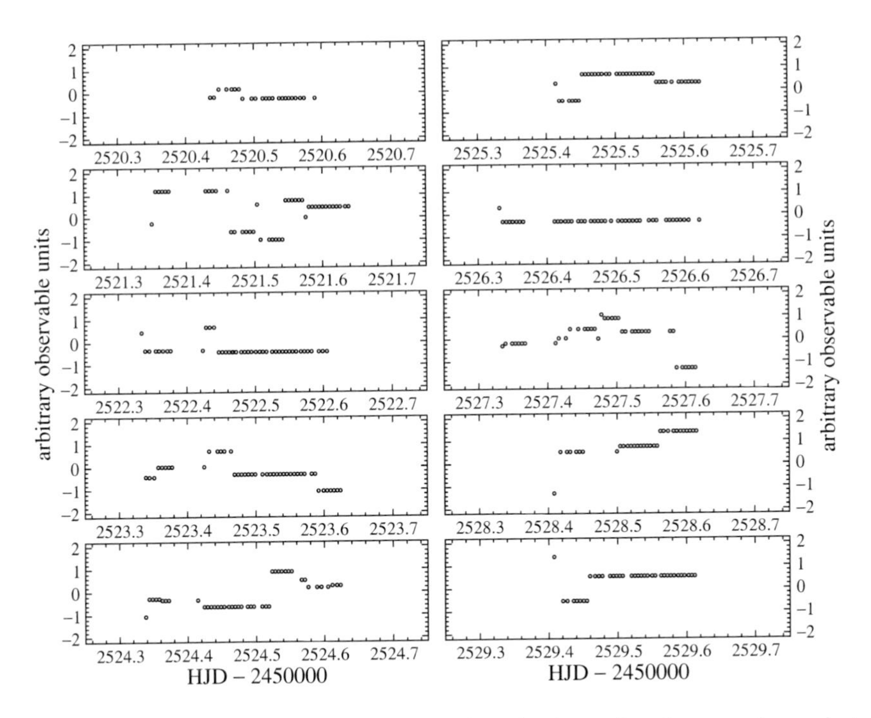

Figure 34: Time series generated by the simulator in the sample project `sim-rndsteps`. The sampling represents the V photometry of IC 4996 # 89. The original observable values are replaced by the simulator.

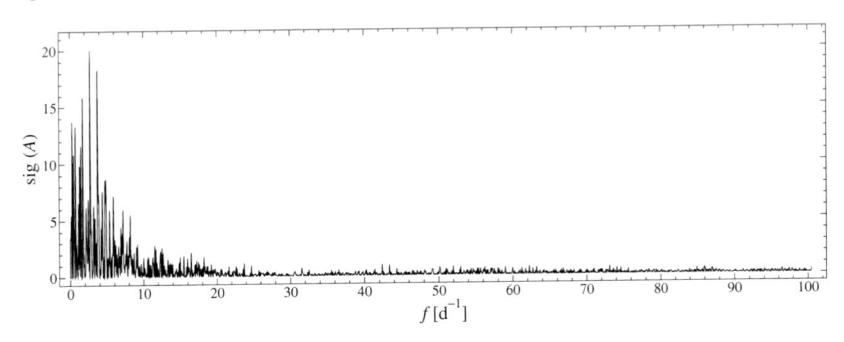

Figure 35: Typical significance spectrum for random steps, based on the sampling of the V photometry of IC 4996 # 89. Each constant in the step function displayed in Fig. 34 contributes a spectral window to this DFT.

in the file `sim-rndsteps.ini` *produces random steps the values of which are distributed according to a Gaussian with standard deviation 0.5. The expected distance in time of consecutive steps 0.07 days. The resulting light curve is displayed in Fig. 34.*

Since the observables are constant between the steps, one may consider each of the corresponding time intervals to contribute a spectral window to the DFT, or significance spectrum, correspondingly. The significance spectrum associated to the light curve in Fig. 34 is displayed in Fig. 35 and represents such a superposition of spectral windows.

13.9. Zero-mean adjustment

The keyword `sim:zeromean` may be used to adjust the mean value of the time series (or a subset) to zero. It is given with two floating-point parameters,

1. the lower time limit, and
2. the upper time limit.

If the lower and upper time limits are both set zero, the mean value of the entire synthetic time series is adjusted to zero. This option was adopted for consistency, but does not provide additional functionality, because a zero-mean correction of the whole data set is performed at every step of the prewhitening cascade by default.

Example. *In the sample project* `sim-zeromean`, SIGSPEC *models the same time series as in the project* `sim-poly` *(p. 67), according to the first part of the file* `sim-zeromean.ini`*:*

```
sim:poly   2520.215 2521.088 4.298 2520.626  0.581
sim:poly   2521.088 2521.679 2.932 2521.443  1.195
sim:poly   2521.679 2522.442 1.092 2522.067  1.063
sim:poly   2522.442 2522.595 5.372 2522.466  0.676
sim:poly   2522.595 2523.351 2.495 2522.682  2.042
sim:poly   2523.351 2523.924 2.839 2523.607  0.221
sim:poly   2523.924 2524.478 8.357 2525.412 -0.899
sim:poly   2524.478 2525.399 2.304 2524.576  1.432
sim:poly   2525.399 2526.107 2.573 2525.721  1.205
sim:poly   2526.107 2526.550 6.350 2526.493  0.031
sim:poly   2526.550 2526.847 4.192 2526.589  2.893
sim:poly   2526.847 2527.616 0.345 2527.652 -0.472
sim:poly   2527.616 2528.264 3.583 2527.783  0.725
sim:poly   2528.264 2528.777 1.246 2528.704  0.610
sim:poly   2528.777 2529.606 3.534 2529.535  1.752
sim:poly   2529.606 2530.242 9.002 2529.694  1.119
```

This block of `sim:poly` *keywords is followed by a corresponding block of* `sim:zeromean` *keywords:*

```
sim:zeromean   2520.215 2521.088
sim:zeromean   2521.088 2521.679
sim:zeromean   2521.679 2522.442
```

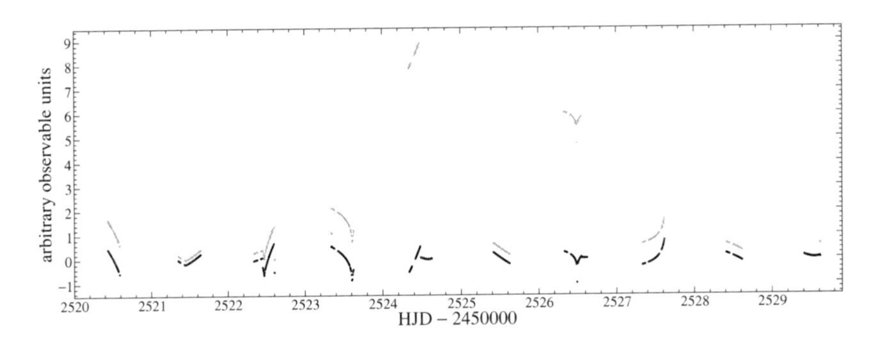

Figure 36: Time series generated by the simulator in the sample project `sim-zeromean`. The sampling represents the V photometry of IC 4996 # 89. First the simulator generates a set of power functions over intervals within the time series (*grey*), then the actual light curve (*black*) is produced by shifting the mean observable for each power function to zero individually.

```
sim:zeromean   2522.442 2522.595
sim:zeromean   2522.595 2523.351
sim:zeromean   2523.351 2523.924
sim:zeromean   2523.924 2524.478
sim:zeromean   2524.478 2525.399
sim:zeromean   2525.399 2526.107
sim:zeromean   2526.107 2526.550
sim:zeromean   2526.550 2526.847
sim:zeromean   2526.847 2527.616
sim:zeromean   2527.616 2528.264
sim:zeromean   2528.264 2528.777
sim:zeromean   2528.777 2529.606
sim:zeromean   2529.606 2530.242
```

This block is responsible for shifting the mean observable to zero for each synthesized power function.

Fig. 36 compares the corresponding light curve with the light curve generated in the project `sim-poly` *(see also Fig. 29). The 16 significant signal components detected by* SigSpec *are of minor interest and not discussed here.*

14. Signal-to-Noise Ratio and Lomb-Scargle Periodogram

As pointed out by Reegen (2007), the SigSpec method represents a tool for an iterative frequency analysis of a zero-mean corrected time series superior to signal-to-noise ratio estimation (Breger et al. 1993) and Lomb-Scargle periodogram (Lomb 1976; Scargle 1982). However, in some situations these alternative methods may be desired or even more reasonable. Namely the Lomb-Scargle periodogram represents the optimum statistical approach to the problem if the mean observable is meaningful rather than set zero arbitrarily. The relations between sig and signal-to-noise ratio or Lomb-Scargle periodogram, respectively, are introduced and discussed by Reegen (2007).

In order to meet a user's requirement of signal-to-noise ratio-based DFT analysis or Lomb-Scargle periodograms as well, the SigSpec software offers the option to perform an analysis relying on amplitude signal-to-noise ratios by providing the keyword `DFT` in the `.ini` file. If this keyword is specified, all SigSpec computations rely on the approximation of sig by the amplitude signal-to-noise ratio according to

$$\mathrm{sig}\,(A) \approx \frac{K \log \mathrm{e}}{4} \frac{A^2}{\langle x^2 \rangle}\,, \tag{27}$$

where K represents the number of time series data, A denotes the Fourier amplitude, and $\langle x^2 \rangle$ refers to the variance of the observable.

Second, the keyword `Lomb` forces SigSpec to evaluate Lomb-Scargle periodograms rather than significance spectra. In this case, the sig is approximated by

$$\mathrm{sig}\,(A) \approx \frac{K \log \mathrm{e}}{4} \frac{P_{\mathrm{LS}}}{\langle x^2 \rangle}\,, \tag{28}$$

where P_{LS} denotes the power level in terms of the Lomb-Scargle periodogram.

Example. *In the sample projects* `DFT` *and* `L-S`*, the input time series represents the V photometry of IC 4996 # 89.*

The file `DFT.ini` *contains a single entry*

```
DFT
```

which forces SigSpec *to rely on the signal-to-noise ratio of DFT amplitudes. The screen output is:*

```
1 freq 3.13205  sig 9.75026  rms 0.00449592  csig 9.75026
2 freq 3.98473  sig 6.80132  rms 0.00422861  csig 6.80083
3 freq 5.40684  sig 5.31609  rms 0.0040257  csig 5.30209
4 freq 17.3677  sig 4.1816  rms 0.00388775  csig 4.14988
```

The file `L-S.ini` *contains a single keyword*

```
Lomb
```

and SigSpec *uses the Lomb-Scargle periodogram rather than sig for all computations. The screen output is:*

```
1 freq 3.13205  sig 9.75026  rms 0.00449592  csig 9.75026
2 freq 3.98472  sig 6.79398  rms 0.00422861  csig 6.7935
3 freq 5.40684  sig 5.31451  rms 0.0040257  csig 5.30033
4 freq 17.3677  sig 4.18161  rms 0.00388775  csig 4.14977
```

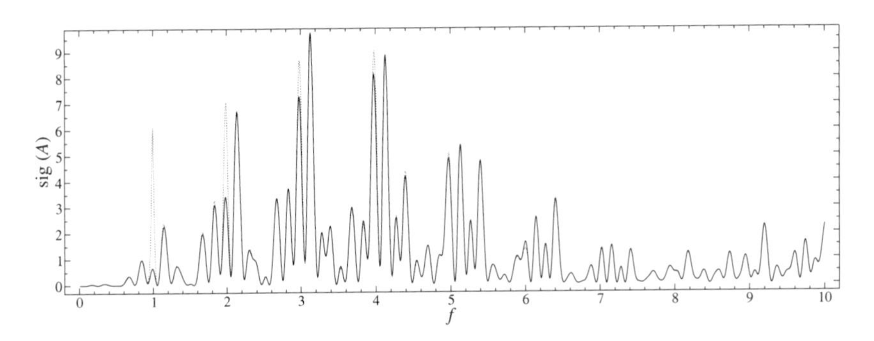

Figure 37: Significance spectrum of the V photometry of IC 4996 # 89 (*grey*) and approximation by the signal-to-noise ratio of DFT amplitudes (*black*).

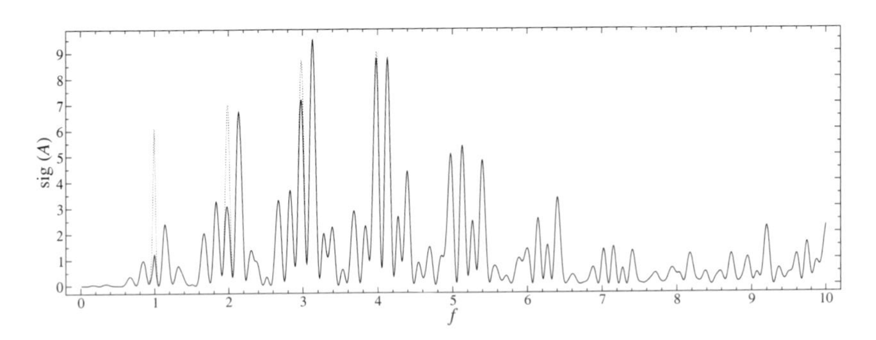

Figure 38: Significance spectrum of the V photometry of IC 4996 # 89 (*grey*) and approximation by the Lomb-Scargle periodogram (*black*).

The significance spectrum of the input time series is compared to the approximations by DFT amplitude signal-to-noise ratio and Lomb-Scargle periodogram in Figs. 37 *and* 38*, respectively.*

A comparison of the two outputs and the screen output of the corresponding sig-based application (Example `SigSpecNative`*, p. 9) reveals slightly different signal components. Especially for the second component the frequency of which is close to an integer multiple of 1 cycle per day and therefore susceptible to alias, the results are different for all three methods. However, the frequencies, amplitudes and phases in the files* `result.dat` *are in good agreement and reflect the numerical uncertainties of the MultiSine fitting procedure only.*

15. Frequently Asked Questions

This section contains questions frequently asked by users familiar to common methods of astronomical time series analysis involving signal-to-noise ratio estimation in power spectra and consecutive prewhitenings. The intention is to clarify the differences between these classical techniques and SIGSPEC from the user's perspective.

15.1. Changing sig in a prewhitening sequence

Given a time series showing two different peaks in the power spectrum, prewhitening of the dominant signal usually does not cause a major change in the height of the secondary peak in the spectrum of the residuals. Why does the corresponding sig change?

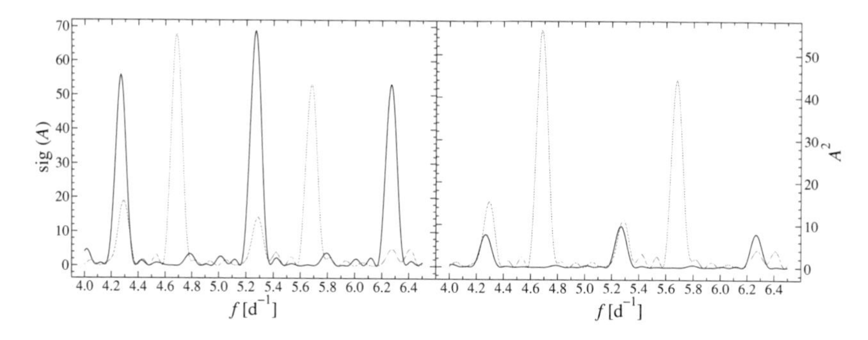

Figure 39: *Grey* graphs: sig (*left*) and power (squared amplitude) spectra (*right*) of a synthetic time series containing two signals plus noise. The sampling represents the V photometry of IC 4996 # 89. *Black* graphs: spectra after subtracting the dominant signal ($f = 4.68573\,\mathrm{d}^{-1}$).

The situation is illustrated in Fig. 39 displaying the sig (*left* panel) and power (*right* panel) spectra (in this sample just squared amplitude) generated by a synthetic time series. It consists of two sinusoidal signals, $f_1 = 4.68573\,\mathrm{d}^{-1}$, $A_1 = 7.27$, and $f_2 = 5.26934\,\mathrm{d}^{-1}$, $A_2 = 3.31$, plus noise with unit rms error. The plots contain a comparison of the initial spectra (*grey*) and the spectra after subtraction of the first signal component. In the right panel, the power associated to the peak at $5.27\,\mathrm{d}^{-1}$ differs only slightly between the two iterations, whereas the corresponding sig in the left panel increases dramatically in the second iteration.

The reason for this behaviour is that the sig refers to the probability of a random time series with the same rms error as the given one to produce a peak like the given one. In the first iteration, the sig calculation is based on the initial time series (rms error 5.84), and in the second iteration, it relies on the residual time series after prewhitening of the peak at f_1, the rms error of which is 2.46. The ratio of rms errors (≈ 2.4) is in agreement with the root ratio of sigs at f_2 in the two iterations (≈ 2.2).

This effect is more prominent for high sigs, because in this case prewhitening causes a major change in the statistical properties of the time series. If a peak with a low sig is prewhitened, the time series is affected marginally, and correspondingly, the sigs of other signals do not change much.

15.2. The effect of binning

Consider a time series representing a signal plus noise. If the data points are grouped into bins, the noise of the binned observables will reduce by the square root of the number of points in each bin. On the other hand, the number of data points the time series consists of reduces by the same amount. Since these two effects cancel each other, the noise level in the power spectrum will be the same for unbinned and binned data. What is the corresponding situation in terms of significance?

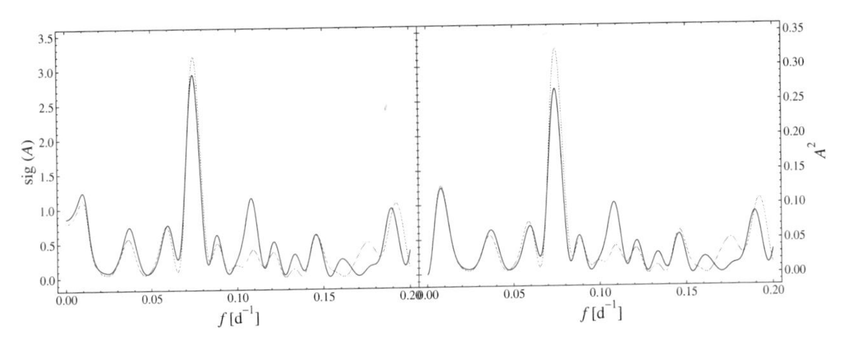

Figure 40: *Grey* graphs: sig (*left*) and power (squared amplitude) spectra (*right*) of a synthetic time series (100 equidistant data points) containing a sinusoidal signal plus noise with a standard deviation of 1. The signal amplitude is 0.5. *Black* graphs: same for time series data grouped into bins of two points. The resulting time series consists of 50 data points.

Fig. 40 contains the significance (*left*) and power (squared amplitude) spectra (*right*) of a synthetic time series containing a sinusoidal signal with a frequency of 0.075832 d^{-1} plus Gaussian noise with a standard deviation of 1. The signal amplitude is 0.5, providing an amplitude signal-to-noise ratio of 5.64. All corresponding plots are displayed in *grey* colour. The *black* graphs represent the spectra generated by a binned version of the time series: each bin contains two data points, and the observable is the arithmetic mean.

In terms of sig as well as amplitude, binning affects neither the peak nor the mean amplitude remarkably: the reduced number of data points would increase the amplitude noise, but this effect is mitigated by the fact that binning reduces the rms residual in the time domain. For a multi-sine signal plus white noise, the number of significant peaks in a given frequency range will hardly be modified by data binning. A considerable change of these sigs by binning is an indication of the noise not being white. A correlation between consecutive measurements in the time series would be a reasonable explanation for such a behaviour.

15.3. Binning of extremely strong signals

If an extremely strong signal is binned, the sig changes, whereas the signal amplitude and the noise level do not. Why?

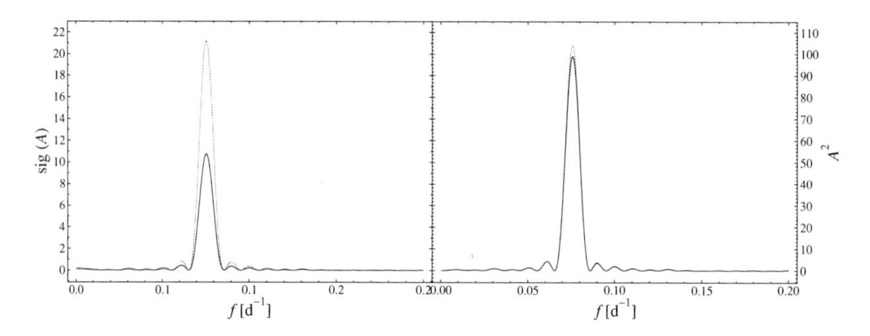

Figure 41: *Grey* graphs: sig (*left*) and power (squared amplitude) spectra (*right*) of a synthetic time series (100 equidistant data points) containing a sinusoidal signal plus noise with a standard deviation of 1. The signal amplitude is 10. *Black* graphs: same for time series data grouped into bins of two points. The resulting time series consists of 50 data points.

Fig. 41 contains the significance (*left*) and power (squared amplitude) spectra (*right*) of a synthetic time series containing a sinusoidal signal with a frequency of 0.075832 d^{-1} plus Gaussian noise with a standard deviation of 1. The signal amplitude of 10 is associated to an ampltiude signal-to-noise ratio of more than 100. All corresponding plots are displayed in *grey* colour. The *black* graphs represent the spectra generated by a binned version of the time series: each bin contains two data points, and the observable is the arithmetic mean.

For both strong and weak signals, binning affects neither the peak nor the mean amplitude remarkably: the reduced number of data points would increase the amplitude noise, but this effect is mitigated by the fact that binning reduces the rms residual in the time domain.

In terms of sig the situation is different: for very strong signals, the peak sig is reduced by binning. Classical techniques prewhiten a peak under consideration and employ the residuals to estimate a noise level. SigSpec does not imply any prewhitening. In the case of a dominant signal plus a tiny scatter, the unbinned and binned data have comparable rms deviations, which are mainly determined by the signal. In the frequency domain, only the reduced number of binned data points comes into play.

Very strong signals let the sig drop to $\approx \frac{1}{N}$ by forming groups of N data points: in Fig. 41, *left* panel, the grey peak is about twice as high as the black peak.

15.4. Linear interpolation: more information?

Consider a time series representing a signal plus noise. Generating additional data points through linear interpolation increases the sig of the signal peak, although the power spectrum remains practically unchanged. This provides the possibility to boost signal sigs artificially, although the amount of information contained by the time series does not increase. Does this make sense?

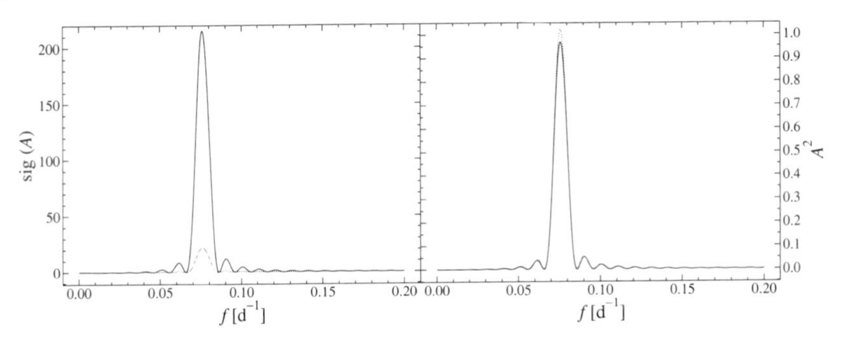

Figure 42: *Grey* graphs: sig (*left*) and power (squared amplitude, logarithmic scale) spectrum (*right*) of a synthetic time series (100 equidistant data points) containing a sinusoidal signal without noise. *Black* graphs: same for a new time series generated by inserting 9 additional linearly interpolated points such that the result is an equidistantly sampled dataset consisting of 991 points.

Fig. 42 displays the sig (*left*) and power (squared amplitude, *right*) spectrum of an equidistantly sampled time series consisting of 100 data points and representing a sinusoidal signal with a frequency of $0.075832\,\mathrm{d}^{-1}$ and an amplitude of 1 in *black* colour. No noise is added. Based on this time series, a new dataset is generated: between each pair of data points, 9 additional, equidistant data points are inserted. The observables are assigned by linear interpolation. The number of data points in this new time series is thus 991. The corresponding spectra are shown in *grey* colour. The longer dataset generates a peak significance that is roughly ten times higher than the initial one, whereas the power spectrum remains practically unchanged. Only the fact that the linear interpolation does not reveal the "true" observables that would be generated by the signal exactly is responsible for a small deviation of the black graph from the grey one.

The explanation for this behaviour is quite similar to the previous section "The effect of binning", p. 81, and correspondingly, the effect is mitigated for very noisy signals. Therefore in practical applications, it will be impossible to enhance the capability of a frequency analysis by artificially introducing new data points.

15.5. Which sig threshold is reasonable?

Occasionally, sigs or sig limits are shifted by $\log \frac{K}{2}$, K denoting the number of time series data points. Which sig threshold is the true one?

In fact both versions are correct, but they apply to different questions. The version without $\log \frac{K}{2}$ refers to the probability that an amplitude level (a peak) *at a given frequency and phase* occurs by chance. The version including $\log \frac{K}{2}$ corresponds to the probability that the highest out of $\frac{K}{2}$ independent peaks occurs by chance.* According to the sampling theorem, the DFT of K data points (a system with K degrees of freedom) produces $\approx \frac{K}{2}$ independent frequencies in Fourier space, if the sampling is equidistant. Although there is no explicit prescription where to find a set of independent frequencies for non-equidistant sampling, the system will still have K degrees of freedom, and the statistical considerations will be essentially the same.

A simple experiment makes the situation clearer: we roll a dice and obtain the result "4". The probability that such an experiment returns at least "4" (i.e. "4", "5" or "6") is, of course, 50 %. This refers to the examination of an individual peak without respect to all the others in the spectrum. If we roll 10 dices, the probability for at least one showing "4" or more is dramatically higher, namely > 99.9 %. This refers to examining the highest out of 10 peaks. The increasing probability of obtaining such a result by chance corresponds to a decreasing significance of the result.

*Note by M. Gruberbauer: The standard SigSpec output corresponds to "the version without $\log \frac{K}{2}$". See Section 5.5.2 of Reegen (2007) for a more detailed explanation of how the significance can be shifted by $\log \frac{K}{2}$ in order to be comparable to the false-alarm probability presented in Scargle (1982).

16. Keywords Reference

This section is a compilation of all keywords accepted by SigSpec. A brief description of arguments and default values is given. The type of argument is provided by either `<int>` or `<double>`, and default values are given in parentheses, e. g. `(2)`. Empty parentheses indicate that there is no default setting.

`antialc:adopt <int> (1)`

number of AntiAlC test iterations adopted for the main prewhitening cascade, p. 48

`antialc:depth <int> (automatic)`

AntiAlC computation depth, p. 48
    parameter: number of iterations used for peak combination testing
    default: $\frac{1}{\sqrt{p_{\rm al}}}$, where $p_{\rm al}$ is the AntiAlC parameter, rounded to the successive integer value

`antialc:par <double> ()`

AntiAlC parameter $p_{\rm al}$: sig limit relative to maximum for the selection of candidate peaks (0 ... use the sig limit `siglimit` instead), p. 47

`antialc:siglimit <double> ()`

significance limit for the AntiAlC candidate peak selection (no significance limit by default; the limit defined by the keyword `siglimit` is used instead), p. 47

`col:obs <int> (2)`

observable column index (unique), starting with 1, p. 13

`col:ssid <int> ()`

subset identifier column index (also multiple), starting with 1, p. 15

`col:time <int> (1)`

time column index (unique), starting with 1, p. 13

`col:weights <int> ()`

weights column index (also multiple), starting with 1, p. 13

`comp <int> ()`

specifies the file indicated by the parameter as comparison dataset, p. 59

`correlograms <int> <int> <int> () ()`

specifications for correlogram files `c#iteration#.dat`, p. 42
parameters:

- correlogram order (maximum index lag), default: half of the number of time series data points,
- number of files to generate (< 0 for all correlogram files, default: no correlogram computation),
- step width (number of iterations) for output.

`csiglimit <double> ()`

lower cumulative sig limit, p. 23

`deftype <target/comp/skip> (target)`

specifies the type of dataset to be assigned to a time series by default, p. 59

`DFT`

forces SigSpec to approximate all sigs by signal-to-noise ratios of DFT amplitudes, p. 78

`diff:comp`

specifies the DFT amplitude spectrum of the comparison datasets to be calculated through a weighted mean of Fourier vectors, p. 60

`diff:compalign`

specifies the DFT amplitude spectrum of the comparison datasets to be calculated as a weighted mean of Fourier amplitudes, p. 60
This setting forces SigSpec to take into account also correlated signal components that lag in phase with respect to each other and the target dataset, respectively.

`diff:off`

switches off the differential significance computation (default), p. 60

`freqspacing <double> ()`

spacing between consecutive frequencies [inverse time units], p. 20

`harmonics <int> ()`

activates the simultaneous analysis of a fundamental plus harmonics (the frequencies of which are integer multiples of the fundamental) the number of which is specified by the parameter, p. 51

`iterations <int> ()`

number of prewhitening iterations, p. 23

`lfreq <double> (0)`

lower frequency limit [inverse time units], p. 18

`Lomb`

forces SIGSPEC to approximate all sigs by the Lomb-Scargle periodogram, p. 78

`mfstart <int> (0)`

index of the first time series input file to apply the MultiFile mode to, p. 57

`mstracks <int> <int> ()`

MultiSine tracks are written to files `m#index#.dat`, where #index# refers to the signal component in the result files. The parameters are

1. the maximum number of iterations for which to write entries to the MultiSine track files, and
2. the step width (number of iterations) for output, p. 37.

The file name may be assigned additional indices for Time-Resolved analysis (p. 42) and/or MultiFile mode (p. 56).

`multifile <int> ()`

activates MultiFile mode, p. 56
parameter: maximum index of time series input files (≤ 0 ... infinite)

`multisine:lock`

forces SigSpec to use the "raw" frequencies, amplitudes, and phases (without MultiSine fitting) for the subsequent analysis, p. 23.

`multisine:newton <double> <double> <double> (0.000001 1 0.000001)`

accuracy parameters for the MultiSine least-squares fits

1. scaling factor for the overall precision of resulting frequencies,
2. degree of dependence of the frequency accuracy on the peak sig,
3. the minimum relative improvement of rms residual between consecutive iterations to continue the fitting process, p. 22.

`multisine:unlock`

forces SigSpec to use the frequencies, amplitudes, and phases improved by MultiSine least-squared fits for the subsequent analysis (default), p. 23.

`nycoef <double> (0.5)`

Nyquist Coefficient (between 0 and 1), p. 19

`nyscan`

Nyquist Coefficients for the specified frequency range (file `nycoef.dat` or `<#multifile#>.nycoef.dat`), p. 20

`osratio <double> (20)`

oversampling ratio, p. 21

`phdist:cart`

generates a Phase Distribution Diagram in three-dimensional cartesian coordinates, p. 35

`phdist:colmodel:lin`

specifies the linear colour model, i. e., phase probability density is used as a colour scale, p. 35

`phdist:colmodel:rank`

specifies the rank colour model, i. e., the rank in an ascending sequence of sock significances is used as a colour scale, p. 35

`phdist:colour <double> <double> <double> <double>`

A set of `phdist:colour` lines defines an RGB path for colourising the Phase Distribution Diagram, p. 35.

parameters:

- red channel (0...255)
- green channel (0...255)
- blue channel (0...255)
- scale

The scale parameter refers directly to probability density of phases in case of `phdist:colmodel:lin`, or to a fractile of probability density on the interval $[0, 1]$ in case of `phdist:colmodel:rank`.

`phdist:cyl`

generates a Phase Distribution Diagram in three-dimensional cylindrical coordinates (default setting), p. 35. The frequency is the height axis, the phase is the azimuth angle, and the radial coordinate refers to the probability density of phase.

`phdist:fill <double> (0)`

specifies a filling factor to compute extra frequencies if the difference of phase PDFs between two adjacent frequencies is too high, p. 35.

parameter: number of additional frequencies per unit probability density (difference between two adjacent frequencies)

`phdist:phases <int> ()`

generates a Phase Distribution Diagram for the sampling of the given time series, p. 34. By default, no Phase Distribution Diagram is computed.

parameter: number of phase angles in the interval $[0, \pi[$, if the maximum probability density is ≤ 1. Between 1 and 2, twice this number is used, and so on. This enhances the visibility of the Phase Distribution Diagram also in frequency and phase regions associated with a very eccentric phase distribution.

`preview <double> ()`

generates a preview, p. 40. Instead of a prewhitening cascade, only one significance spectrum is computed. All local maxima above the specified significance limit are written to a file `preview.dat`. By default, no preview is computed.
parameter: significance limit

`profile`

SigSpec generates a file `profile.dat` containing the sampling profile for the given time series, p. 30. By default, the file `profile.dat` is not generated.

This keyword is ignored in MultiFile mode, where sampling profiles are calculated and written to files whenever required by the program. See "MultiFile Mode", p. 56 for further information.

`residuals <int> <int> ()`

output files containing residual time series (only `residuals.dat` for the residuals after prewhitening all significant compontents by default). The parameters are

1. the maximum number of iterations (files `t#iteration#.dat`), and

2. the step width (number of iterations) for output, p. 27.

The file name may be assigned additional indices for Time-Resolved analysis (p. 42) and/or MultiFile mode (p. 56).

`results <int> <int> ()`

output files containing a list of significant signal components. The default setting is to produce only a file `result.dat` for the final list. The parameters are

1. the maximum number of iterations for which to write additional result files `r#iteration#.dat`, and

2. the step width (number of iterations) for output, p. 28.

The file name may be assigned additional indices for Time-Resolved analysis (p. 42) and/or MultiFile mode (p. 56).

`siglimit <double> (5)`

lower sig limit (0 to deactivate), p. 23

`sim:add`

add synthetic data to given observable, p. 63

`sim:exp <double> <double> <double> <double> <double> ()`

exponential trend, p. 69
parameters:

- lower time limit [time units]
- upper time limit [time units]
- scale
- time zeropoint [time units]
- exponent

`sim:off`

deactivate simulator (default), p. 63

`sim:poly <double> <double> <double> <double> <double> ()`

polynomial trend, p. 67
parameters:

- lower time limit [time units]
- upper time limit [time units]
- scale
- time zeropoint [time units]
- exponent

full polynomial by multiple declaration with different scales, time zeropoints, and exponents

`sim:replace`

replace given observable by synthetic data, p. 63

sim:rndsteps <double> <double> <double> <double> ()

random steps, p. 74
parameters:

- lower time limit [time units]
- upper time limit [time units]
- standard deviation for Gaussian distribution of (constant) step values
- expected time range for Poisson distribution of steps [time units]

sim:serial <double> <double> <double> <double> <double> <double> ()

serially correlated noise, p. 70
parameters:

- lower time limit [time units]
- upper time limit [time units]
- scale for standard deviation
- time zeropoint for polynomial trend of standard deviation [time units]
- exponent for polynomial trend of standard deviation
- serial correlation coefficient

full polynomial by multiple declaration with different scales, time zeropoints, and exponents

sim:signal <double> <double> <double> <double> <double> ()

sinusoidal signal, p. 64
parameters:

- lower time limit [time units]
- upper time limit [time units]
- amplitude
- time zeropoint [time units]
- frequency [inverse time units]

`sim:temporal <double> <double> <double> <double> <double> <double> ()`

temporally correlated noise, p. 72
parameters:

- lower time limit [time units]
- upper time limit [time units]
- scale for standard deviation
- time zeropoint for polynomial trend of standard deviation [time units]
- exponent for polynomial trend of standard deviation
- temporal correlation coefficient

full polynomial by multiple declaration with different scales, time zeropoints, and exponents

`sim:zeromean <double> <double> ()`

zero-mean adjustment, p. 76
parameters:

- lower time limit [time units]
- upper time limit [time units]

`skip <int> ()`

forces SigSpec to skip the file indicated by the parameter, p. 59

`sock:cart`

generates a Sock Diagram in three-dimensional cartesian coordinates, p. 32

`sock:colmodel:lin`

specifies the linear colour model, i. e., sock significance is used as a colour scale, p. 32

`sock:colmodel:rank`

specifies the rank colour model, i. e., the rank in an ascending sequence of sock significances is used as a colour scale, p. 32

`sock:colour <double> <double> <double> <double>`

A set of `sock:colour` lines defines an RGB path for colourising the Sock Diagram, p. 33.

parameters:

- red channel (0...255)
- green channel (0...255)
- blue channel (0...255)
- scale

For the linear colour model selected by the keyword `sock:colmodel:lin`, the scale parameter refers directly to sock significance. If the rank colour model is selected (`sock:colmodel:rank`), it refers to a fractile of sock significance on the interval $[0, 1]$.

`sock:cyl`

generates a Sock Diagram in three-dimensional cylindrical coordinates (default setting), p. 32. The frequency is the height axis, the phase is the azimuth angle, and the radial coordinate refers to the sock significance.

`sock:fill <double> (0)`

specifies a filling factor to compute extra frequencies if the sock significance difference between two adjacent frequencies is too high, p. 32.

parameter: number of additional frequencies per unit sig (difference between two adjacent frequencies)

`sock:phases <int> ()`

generates a Sock Diagram for the sampling of the given time series, p. 31. By default, no Sock Diagram is computed.

parameter: number of phase angles in the interval $[0, \pi[$, if the maximum sock significance is ≤ 1. Between 1 and 2, twice this number is used, and so on. This enhances the visibility of the Sock Diagram also in frequency and phase regions associated with a high sock significance.

`spectra <int> <int> ()`

output files containing spectra (only `s000000.dat` for the spectrum of the initial time series and `resspec.dat` for the spectrum of the residuals after prewhitening all significant compontents by default). The parameters are

1. the maximum number of iterations (files `s#iteration#.dat`), and
2. the step width (number of iterations) for output, p. 26.

The file name may be assigned additional indices for Time-Resolved analysis (p. 42) and/or MultiFile mode (p. 56).

`target <int> ()`

specifies the file indicated by the parameter as target dataset, p. 59

`timeres:range <double> ()`

subset interval width [time units], p. 42

`timeres:step <double> ()`

step width between subset centres [time units], p. 42

`timeres:w:cos <double> <double> ()`

cosine weights, p. 44
parameters:

- frequency [inverse time units]
- phase [rad]

`timeres:w:cosp <double> <double> <double> ()`

weights according to the power of a cosine, p. 44
parameters:

- frequency [inverse time units]
- phase [rad]
- exponent

`timeres:w:damp <double> ()`

exponential damping, p. 44
parameter: width [time units]

`timeres:w:exp <double> ()`

exponential weights, p. 44
parameter: width [time units]

`timeres:w:gauss <double> ()`

Gaussian weights, p. 44
parameter: standard deviation [time units]

`timeres:w:ipow <double> ()`

inverse power weights, p. 44
parameter: exponent

`timeres:w:none`

unweighted moving averages, i. e. a rectangular filter, p. 44

`ufreq <double> ()`

upper frequency limit [inverse time units], p. 19

`win`

SigSpec generates a file `win.dat` containing the spectral window for the given time series. By default, the file `win.dat` is not generated, p. 29.

17. Online availability

The ANSI-C code is available online at `http://www.sigspec.org`. For further information, please contact P. Reegen, `peter.reegen@univie.ac.at`.*

Acknowledgments. PR received financial support from the Fonds zur Förderung der wissenschaftlichen Forschung (FWF, projects P14546-PHY, P17580-N2) and the BM:BWK (project COROT). Furthermore, it is a pleasure to thank T. Appourchaux (IAS, Orsay), A. Baglin (Obs. de Paris, Meudon),

*Please contact Michael Gruberbauer, `mgruberbauer@ap.smu.ca`.

T. Boehm (Obs. M.-P., Toulouse), M. Breger, R. Dvorak, M. G. Firneis, D. Frast (Univ. of Vienna), R. Garrido (Inst. Astrof. Andalucia, Granada), M. Gruberbauer (Univ. of Vienna), D. B. Guenther (St. Mary's Univ., Halifax), M. Hareter, D. Huber, T. Kallinger (Univ. of Vienna), R. Kuschnig (UBC, Vancouver), S. Marchenko (Western Kentucky Univ., Bowling Green, KY), M. Masser (Univ. of Vienna), J. M. Matthews (UBC, Vancouver), E. Michel (Obs. de Paris, Meudon), A. F. J. Moffat (Univ. de Montreal), E. Paunzen, D. Punz (Univ. of Vienna), V. Ripepi (INAF, Naples), S. M. Rucinski (D. Dunlap Obs., Toronto), T. A. Ryabchikova (Inst. Astrpn. RAS, Moscow), D. Sasselov (Harvard-Smithsonian Center, Cambridge, MA), S. Schraml (Univ. of Technology, Vienna), G. A. Wade (Royal Military College, Kingston), G. A. H. Walker (UBC, Vancouver), W. W. Weiss, and K. Zwintz (Univ. of Vienna) for valuable discussion and support with extensive software tests. I acknowledge the anonymous referee for a detailed examination of both this publication and the corresponding software, as well as for the constructive feedback that helped to improve the overall quality a lot. Finally, I address my very special thanks to J. D. Scargle for his valuable support.

References

Breger, M., Stich, J., Garrido, R., et al. 1993, A&A, 271, 482
Breger, M., Rucinski, S. M., Reegen, P. 2007, AJ, 134, 1994
Kallinger, T., Reegen, P., Weiss, W. W. 2008, A&A, 481, 571
Kjeldsen, H., Bedding, T. R., Viskum, M., Frandsen, S. 1995, AJ, 109, 1313
Lomb, N. R. 1976, ApSS, 39, 447
Reegen, P. 2005, in *The A-Star Puzzle*, Proceedings of IAUS 224, eds. J. Zverko, J. Ziznovsky, S.J. Adelman, W.W. Weiss (Cambridge: Cambridge Univ. Press), p. 791
Reegen, P. 2007, A&A, 467, 1353
Scargle, J. D. 1982, ApJ, 263, 835
Strassmeier, K. G., Boyd, L. J., Epand, D. H., Granzer, T. 1997, PASP, 109, 697
Zwintz, K., Marconi, M., Kallinger, T., Weiss, W. W. 2004, in *The A-Star Puzzle*, Proceedings of IAUS 224, eds. J. Zverko, J. Ziznovsky, S. J. Adelman, W. W. Weiss (Cambridge: Cambridge Univ. Press), p. 353
Zwintz, K., Weiss, W. W. 2006, A&A, 457, 237

Comm. in Asteroseismology
Volume 163, 2011

Cinderella User's Manual

P. Reegen

Institut für Astronomie, Türkenschanzstraße 17, 1180 Vienna, Austria

Abstract

CINDERELLA is a software solution for the quantitative comparison of time series in the frequency domain. It assigns probabilities to coincident peaks in the DFT amplidude spectra of the datasets under consideration. Two different modes are available. In conditional mode, CINDERELLA examines target and comparison datasets on the assumption that the latter contain artifacts only, returning the conditional probability of a target signal, although there is a coincident signal in the comparison data within the frequency resolution. In composed mode, the probability of coincident signal components in both target and comparison data is evaluated. CINDERELLA permits to examine multiple target and comparison datasets at once.

1. What is CINDERELLA?

CINDERELLA is an abbreviation of "**C**omparison of **INDE**pendent **REL**ative **L**east-squares **A**mplitudes". It provides a quantitative comparison between the DFT amplitude spectra of time-resolved astronomical measurements.

The SIGSPEC technique (Reegen 2005, 2007) allows to determine probabilities for coincident peaks in the DFT amplitude spectra of different datasets quantitatively and in a statistically unbiased way. The theoretical background of this procedure is introduced by Reegen et al. (2008).

CINDERELLA uses the standard output of the program SIGSPEC, which represents the results of a cascade of consecutive prewhitenings employing least-squares fits to obtain amplitudes and phases of signal components. Following the SIGSPEC terminology, CINDERELLA returns a spectral significances (hereafter abbreviated by 'sig') rather than a probability. This manual uses cross-references to SIGSPEC frequently. In these cases, the reader is referred to the SIGSPEC manual (Reegen 2011).

2. Projects

CINDERELLA is called by the command line

```
Cinderella <project>
```

where <project> is the name (or path, if desired) of the CINDERELLA project. The project structure is strictly consistent with SIGSPEC.

Before running the program, the user has to provide

1. at least two time series input files consistent with the SIGSPEC MultiFile mode (see SIGSPEC manual, p. 56, and "Time series input files", p. 100), and

2. a directory <project> containing the SIGSPEC result files corresponding to the time series input files (see "SIGSPEC result files", p. 101).

Furthermore, the user may pass a set of specifications to CINDERELLA by means of a file <project>.cnd (see "The .cnd file", p. 105). This file is expected in the same folder as the time series input files and the project directory. For specifications not given by the user, defaults are used.

CINDERELLA is designed to answer two different questions, depending on the problem it is applied to. By default, the program runs through both modes simultaneously and provides both conditional (p. 108) and composed sigs (p. 114).

3. Input

3.1. Time series input files

CINDERELLA expects at least two time series input files named according to #index#.<project>.dat. In this context, #index# is a placeholder for a six-digit index of the file. Note that only files with consecutive indices are appropriately recognised by CINDERELLA. The conventions are compatible with the SIGSPEC MultiFile mode (see SIGSPEC manual, p. 56), whence the most convenient preparation of data for CINDERELLA is the SIGSPEC MultiFile computation.

The only restrictions to the format of the time series input files are that the number of items per row has to be constant for all rows in the file and that the columns have to be seperated by at least one whitespace character or tab. Dataset entries need not to be numeric, except for the columns specified as time, observable, and weights. The conventions for specifying these three column types are consistent with SIGSPEC. See SIGSPEC manual, p. 13 for details.

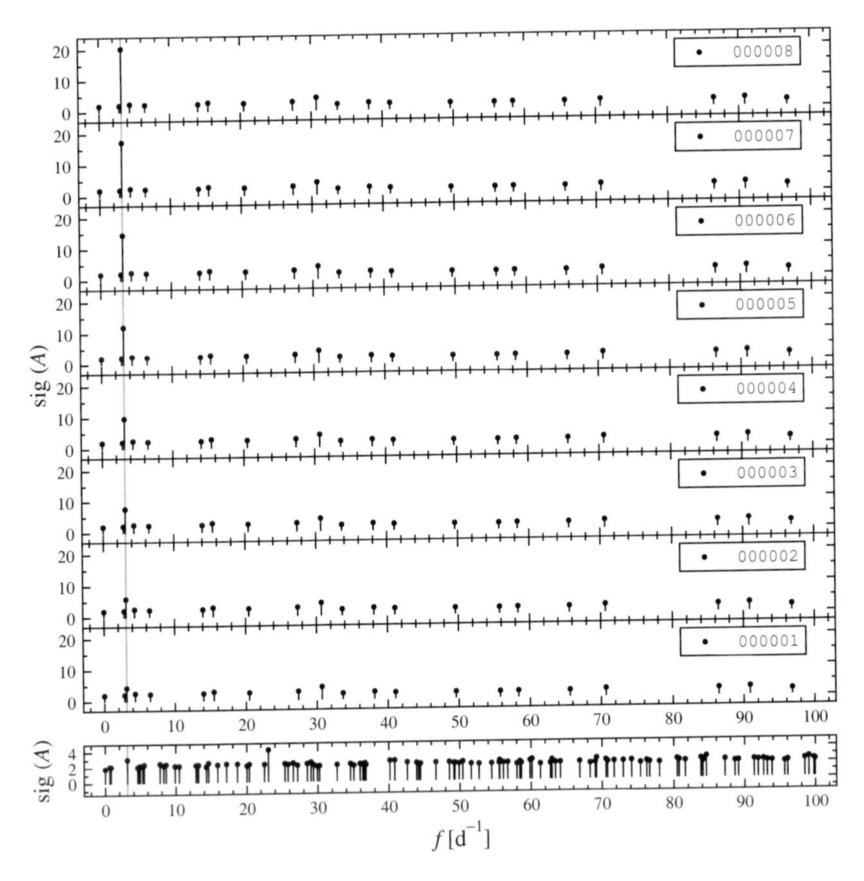

Figure 1: SigSpec result files used as input for the sample project `CinderellaNative`. The bottom panel refers to the file `CinderellaNative/000000.result.dat`, which is used as the comparison dataset. Above, the files `CinderellaNative/000001.result.dat` to `CinderellaNative/000008.result.dat` are displayed from bottom to top. The underlying time series represent Gaussian noise plus a single sinusoidal signal at a frequency of 3.125 d^{-1} (grey line) with different amplitudes.

3.2. SigSpec result files

The SigSpec result files `#index#.result.dat` are located in the project directory and contain a list of all sig maxima associated to each of the input time series `#index#.<project>.dat`. A SigSpec result file consists of seven columns. A full description is provided by the SigSpec manual, p. 28. Cinderella uses only five columns:

- the frequency [inverse time units] (column 1),
- the sig (column 2),
- the amplitude [units of observable] (column 3),
- rms scatter of the time series before prewhitening (column 5), and
- point-to-point scatter of the time series before prewhitening (column 6).

The last line of a result file contains only two non-zero values in columns 5 and 6. These represent the rms and point-to-point scatter of the time series after the last prewhitening step, correspondingly, and are also used by CINDERELLA.

Example. *The sample project* `CinderellaNative` *provides a run without any additional options by typing* `Cinderella CinderellaNative`. *The files* `000000.CinderellaNative.dat`, ..., `000008.CinderellaNative.dat` *are the same as* `000038.diffsig.dat` *to* `000046.diffsig.dat` *in the* SIGSPEC *example* `diffsig` *(*SIGSPEC *manual, p. 60), correspondingly.*

The SIGSPEC *result files* `000000.result.dat` *to* `000008.result.dat` *are provided in the project directory* `CinderellaNative`. *They contain all signal components found with sig > 2 and are displayed in Fig.* 1.

The screen output at runtime starts with a standard header.

```
 CCCCC  ii            dd                      ll  ll
CC   CC               dd                      ll  ll
CC      ii n nnn   ddddd  eeee  r rrr   eeee  ll  ll   aaaa
CC      ii nn  nn dd  dd ee  ee rr  rr ee  ee ll  ll  aa  aa
CC      ii nn  nn dd  dd ee  ee rr     ee  ee ll  ll      aa
CC      ii nn  nn dd  dd eeeeee rr     eeeeee ll  ll   aaaaa
CC      ii nn  nn dd  dd ee     rr     ee     ll  ll  aa  aa
CC   CC ii nn  nn dd  dd ee  ee rr     ee  ee ll  ll  aa  aa
 CCCCC  ii nn  nn  ddd d  eeee  rr      eeee   ll  ll  aaa a


Comparison of INDEpendent RELative Least-squares Amplitudes
Version 1.0
************************************************************
by Piet Reegen
Institute of Astronomy
University of Vienna
Tuerkenschanzstrasse 17
1180 Vienna, Austria
Release date: April 29, 2008
```

The program starts with processing the command line, checking if a present directory `CinderellaNative` *is present, and searching for a configuration file* `CinderellaNative.cnd` *(see "The* `.cnd` *file", p. 105). Since there is no such file present,* CINDERELLA *produces a warning message.*

```
*** start ***************************************************

Checking availability of project directory CinderellaNative...
project directory CinderellaNative ok.

Warning: CndFile_LoadCnd 001
         Failed to open .cnd file.
```

Now CINDERELLA *counts the time series input files and checks for corresponding* SIGSPEC *result files.*

```
*** MultiFile count ***************************************

Number of files                           9
```

The next step is to count the rows in each SIGSPEC *result file.*

```
*** count rows in SigSpec result files ********************

CinderellaNative/000000.result.dat:    106 rows
CinderellaNative/000001.result.dat:     21 rows
CinderellaNative/000002.result.dat:     21 rows
CinderellaNative/000003.result.dat:     21 rows
CinderellaNative/000004.result.dat:     21 rows
CinderellaNative/000005.result.dat:     21 rows
CinderellaNative/000006.result.dat:     21 rows
CinderellaNative/000007.result.dat:     21 rows
CinderellaNative/000008.result.dat:     21 rows
```

Before reading the input files, CINDERELLA *performs the assignment of* `target`, `comp` *and* `skip` *tags to the datasets (see "Dataset Types", p. 106). Since there is no file* `CinderellaNative.cnd` *available, the defaults are used: the first file,* `000000.CinderellaNative.dat` *is considered a comparison dataset, the eight remaining files are interpreted as targets.*

```
*** dataset type assignment *******************************


Warning: CndFile_Cind 001
         Failed to open .cnd file.
         Assigning default types.

000000.CinderellaNative.dat: comparison (default)
000001.CinderellaNative.dat: target (default)
000002.CinderellaNative.dat: target (default)
000003.CinderellaNative.dat: target (default)
000004.CinderellaNative.dat: target (default)
000005.CinderellaNative.dat: target (default)
000006.CinderellaNative.dat: target (default)
000007.CinderellaNative.dat: target (default)
000008.CinderellaNative.dat: target (default)

number of target datasets                 8
number of comparison datasets             1
number of datasets to skip                0
```

The time series and SIGSPEC result files are read, and CINDERELLA displays the frequency resolution and the mean observable for each time series.

```
*** read input files ***************************************

000000.CinderellaNative.dat: Rayleigh resolution 0.1089880382935977
000000.CinderellaNative.dat: mean observable      0.0074422789800987
CinderellaNative/000000.result.dat
000001.CinderellaNative.dat: Rayleigh resolution 0.0914470160931467
000001.CinderellaNative.dat: mean observable     -0.0704869463068650
CinderellaNative/000001.result.dat
000002.CinderellaNative.dat: Rayleigh resolution 0.0914470160931467
000002.CinderellaNative.dat: mean observable     -0.0875484339175066
CinderellaNative/000002.result.dat
000003.CinderellaNative.dat: Rayleigh resolution 0.0914470160931467
000003.CinderellaNative.dat: mean observable     -0.1046099215281657
CinderellaNative/000003.result.dat
000004.CinderellaNative.dat: Rayleigh resolution 0.0914470160931467
000004.CinderellaNative.dat: mean observable     -0.1216714091388107
CinderellaNative/000004.result.dat
000005.CinderellaNative.dat: Rayleigh resolution 0.0914470160931467
000005.CinderellaNative.dat: mean observable     -0.1387328967494604
CinderellaNative/000005.result.dat
000006.CinderellaNative.dat: Rayleigh resolution 0.0914470160931467
000006.CinderellaNative.dat: mean observable     -0.1557943843601256
CinderellaNative/000006.result.dat
000007.CinderellaNative.dat: Rayleigh resolution 0.0914470160931467
000007.CinderellaNative.dat: mean observable     -0.1728558719707690
CinderellaNative/000007.result.dat
000008.CinderellaNative.dat: Rayleigh resolution 0.0914470160931467
000008.CinderellaNative.dat: mean observable     -0.1899173595814251
CinderellaNative/000008.result.dat
```

The amplitudes in the SIGSPEC result file of the comparison dataset (file `CinderellaNative/000000.result.dat` *have to be transformed in order to be comparable to the target datasets (see "Amplitude transformation", p. 112). CINDERELLA uses the rms residual as a measure for this transformation. This is the default setting.*

```
*** amplitude transformation ********************************

by rms error: file                     0
```

The core of CINDERELLA consists of three different analyses. The first part of these is the computation of conditional sigs for each pair of target vs. comparison datasets. In this case we have eight target datasets and one comparison dataset, which results in eight operations. In general, the total number of operations performed in this step is the product of the number of target datasets times the number of comparison datasets. Detailed information on the computation of pairwise conditional sigs is found in "Single-comparison output" (p. 109).

```
*** pairwise Cinderella analysis *************************

    1 vs.      0: conditional sig
    2 vs.      0: conditional sig
    3 vs.      0: conditional sig
    4 vs.      0: conditional sig
    5 vs.      0: conditional sig
    6 vs.      0: conditional sig
    7 vs.      0: conditional sig
    8 vs.      0: conditional sig
```

The second part of the CINDERELLA *analysis is a computation of mean conditional sigs for each target dataset over all comparison datasets (see "Multi-comparison output", p. 109). Since there is only one comparison dataset available in this example, the output of this procedure is the same as of the previous pairwise analysis. Finally,* CINDERELLA *evaluates composed sigs for each target dataset. This calculation is applied to the "raw"* SIGSPEC *results and also to the conditional sigs obtained by the previous operation. Details on the computation of composed sigs are found in "Composed Mode" (p. 114).*

```
*** total Cinderella analysis *****************************

    1: conditional sig
    2: conditional sig
    3: conditional sig
    4: conditional sig
    5: conditional sig
    6: conditional sig
    7: conditional sig
    8: conditional sig
composed sig
composed sig of conditional sigs
```

On exit, CINDERELLA *displays a goodbye message.*

```
Finished.

************************************************************

Thank you for using Cinderella!
Questions or comments?
Please contact Piet Reegen (reegen@astro.univie.ac.at)
Bye!
```

The CINDERELLA *output for this example is discussed in the subsequent chapters.*

3.3. The .cnd file

An optional file `<project>.cnd` consists of a set of keywords and arguments defining project-specific parameters for CINDERELLA. If this file is not present in the same folder as the time series input files, CINDERELLA uses a set of default parameters.

Caution: CINDERELLA **demands a carriage-return character at the end of the file** `<project>.cnd`**, otherwise the program hangs!**

Lines in the .cnd file starting with a # character are ignored by CINDERELLA. This provides the possibility to write comments into the file.

4. Indexing

By default, CINDERELLA expects the file index of time series and SIGSPEC results to start at zero. If the user wants the program to start at a different index, the keyword `mfstart` may be given in the .cnd file. This keyword is followed by an integer representing the desired start index. Furthermore, the software takes into account as many files with consecutive indices as available. The number of indices to process may be restricted by means of the keyword `multifile`, followed by an integer representing the last index to take into account. The use of the keywords `mfstart` and `multifile` is strictly consistent with the SIGSPEC conventions (SIGSPEC manual, p. 56).

Example. *The sample project* `index` *contains the same input as the project* `CinderellaNative` *(p. 102), but both the time series input files and the* SIGSPEC *result files are now enumerated from* 004847 *to* 004855 *rather than from* 000000 *to* 000008. *The file* `index.cnd` *consists of a single line,*

```
mfstart 4847
```

Since the keyword `multifile` *is not specified in the file* `index.cnd`, CINDERELLA *uses all input files available through consecutive indices.*

The computations are entirely the same as for `CinderellaNative`, *but all indices are consistently incorporated by* CINDERELLA. *E. g., the screen output for the single-comparison computations is now:*

```
*** pairwise Cinderella analysis ***************************

  4848 vs.   4847: conditional sig
  4849 vs.   4847: conditional sig
  4850 vs.   4847: conditional sig
  4851 vs.   4847: conditional sig
  4852 vs.   4847: conditional sig
  4853 vs.   4847: conditional sig
  4854 vs.   4847: conditional sig
  4855 vs.   4847: conditional sig
```

5. Dataset Types

In order to avoid unnessecarily high computational effort and redundant output, if comparing all possible pairs of time series input files, CINDERELLA provides the possibility to specify which pairs of target/comparison datasets to take into account. Moreover, the user has the opportunity to identify datasets to be ignored.

CINDERELLA will produce one so-called single-comparison output file (see "Single-comparison output", p. 109) for each target-comparison pair. If there is more than one comparison dataset available, additional files are generated for each target. They contain summaries concerning all comparison datasets examined for the target (see "Multi-comparison output", p. 109) and "Output for composed mode", p. 114).

Contrary to the file nomenclature, the six-digit format is not required for file indices specified in the `.cnd` file.

5.1. Target datasets

The keyword `target` in the `.cnd` file is used for the specification of a target dataset. The keyword is followed by an integer value referring to the six-digit index of the desired time series input file. Multiple declarations of `target` are supported. If no `.cnd` file is available, CINDERELLA uses the first time series input file (i. e. the one with the start index) as the only target dataset.

5.2. Comparison datasets

The keyword `comp` in the `.cnd` file is used for the specification of a comparison dataset. The keyword is followed by an integer value referring to the six-digit index of the desired time series input file. Contrary to the file nomenclature, the six-digit format is not required for file indices specified in the `.cnd` file. If no `.cnd` file is available, CINDERELLA uses all time series input files – except for the first one, which is considered target data – as comparison datasets.

5.3. Datasets to ignore

The keyword `skip` in the `.cnd` file is used for the specification of a dataset not to be taken into account for computation. The keyword is followed by an integer value referring to the six-digit index of the desired time series input file. Contrary to the file nomenclature, the six-digit format is not required for file indices specified in the `.cnd` file.

5.4. Default type

To enhance the convenience for the user, not all files need to be specified by the keywords `target`, `comp` and `skip`. The keyword `deftype` may be used to assign a default dataset type.

1. Use `deftype target` to assign the target attribute by default. If no `deftype` keyword is provided, this setting is activated.
2. Use `deftype comp` to assign the comp attribute by default.
3. Use `deftype skip` to assign the skip attribute by default.

Example. *The sample project* `types` *contains the same input as the project* `CinderellaNative` *(p. 102), and the file* `types.cnd` *contains the two lines*

```
deftype target
comp 0
```

This reproduces the default assignment of data types, and CINDERELLA *performs the same calculations as for the project* `CinderellaNative`. *The only difference is the screen output:*

```
*** dataset type assignment ********************************

000000.types.dat: comparison
000001.types.dat: target
000002.types.dat: target
000003.types.dat: target
000004.types.dat: target
000005.types.dat: target
000006.types.dat: target
000007.types.dat: target
000008.types.dat: target
```

The fact that `(default)` *is not attached to the file list indicates that* CINDERELLA *uses the specifications given in the file* `types.cnd` *rather than the standard settings applied to the project* `CinderellaNative`.

6. Conditional Mode

The conditional sig is a measure of the probability that a signal component in a target star occurs, although a coincident signal component is found in a comparison star or sky background. It provides an answer to the question, "What is the probability that a signal component with given amplitude and sig in the target data is not due to the same process that produces a coincident signal component with given amplitude and sig in the comparison data?"

The conditional CINDERELLA mode is comparable to the differential sig (SIGSPEC manual, p. 58), although the numerical results are not the same.

- For the differential mode of SIGSPEC, the full spectral information is available. Thus SIGSPEC handles the DFT spectra as continuous functions. CINDERELLA accesses only a list of peaks detected by SIGSPEC.

Deviations of corresponding peak frequencies in comparison and target spectra cannot be handled as accurately as in the case of differential sig computation.

- The differential mode of SigSpec compares power integrals over the entire frequency range under consideration for the transformation of amplitudes from comparison into target data. Since Cinderella deals with a list of frequencies rather than the entire spectra, different strategies to transform amplitudes have to be employed. See "Amplitude transformation", p. 112.

Single-comparison output

The single-comparison output files contain three columns:

1. target frequency [inverse time units],
2. conditional sig,
3. conditional csig.

Each file refers to the analysis of a single target-comparison dataset pair. Correspondingly, two six-digit indices `#target#`, `#comparison#` are used to form the filenames, `#target#.cd.#comparison#.dat` (conditional mode) and `#target#.cd.#comparison#.dat` (composed mode).

Example. *The output file* `000004.cd.000026.dat` *contains conditional sigs for* `000004.<project>.dat` *as target data and* `000026.<project>.dat` *as comparison data.*

Example. *In the sample* `CinderellaNative`*, there are 8 single-comparison output files:* `000001.cd.000000.dat` *to* `000008.cd.000000.dat`*. Since there is only one comparison dataset available, these files are redundant, because* `#target#.cd.dat` = `#target#.cd.000000.dat`.

Multi-comparison output

For each target dataset, a multi-comparison output file `#target#.cd.dat` is generated for each target dataset. The three columns represent

1. target frequency [inverse time units],
2. mean conditional sig for all comparison datasets,
3. mean conditional csig for all comparison datasets.

If there is only one comparison dataset available, the single-comparison and multi-comparison output files are identical.

Example. *The sample* `CinderellaNative` *contains 8 multi-comparison output files in the project directory:* `000001.cd.dat` *to* `000008.cd.dat`.

6.1. Candidate selection

For each target dataset, CINDERELLA scans all comparison datasets, searching for coincident signal components. A pair of signal components is considered coincident, if the frequencies match to an accuracy that may be specified by the user, who may also define what CINDERELLA shall do if a comparison dataset does not contain a match for a given target frequency.

If more than one coincident frequency associated to a given target frequency is found, CINDERELLA chooses the candidate with the lowest conditional or composed sig in order to obtain the most conservative solution.

Frequency resolution

There are mainly two interpretations of the frequency resolution. It is either calculated as the inverse time interval width (Rayleigh frequency resolution),

$$\delta f := \frac{1}{T}\,, \tag{1}$$

or as

$$\delta f := \frac{1}{T\sqrt{\mathrm{sig}\,(A)}}\,, \tag{2}$$

where $\mathrm{sig}\,(A)$ denotes the sig of an amplitude A. This definition is called Kallinger resolution (Kallinger, Reegen & Weiss 2008).

In order to enhance the flexibility of CINDERELLA, the frequency resolution is computed as

$$\delta f := \frac{1}{T\sqrt{\mathrm{sig}\,(A)^{\tau}}}\,, \tag{3}$$

where the floating-point number $\tau \in [0,1]$ may be defined using the keyword `tol` in the `.cnd` file. The special value $\tau = 0$ transforms eq. 3 into eq. 1, setting $\tau = 1$ provides eq. 2. The default value is $\tau = 0$.

CINDERELLA checks for frequencies in the comparison datasets are within the frequency resolution around each frequency in the target dataset.

Example. *The sample* `cand` *contains the same input as* `CinderellaNative` *(p. 102), and the file* `cand.cnd` *contains the line*

```
tol 2
```

The frequency tolerance parameter is increased compared to the default value 0, which means that the intervals taken into account to search for corresponding signal components are tendentially narrower. This setting is for demonstration only; in normal applications, only frequency tolerance parameters ranging from 0 to 1 will make sense.

The effect of this modification is visible, e. g., comparing the output files `000001.cd.000000.dat` *of the project* `CinderellaNative` *to the project* `cand`. *In the project* `CinderellaNative`, *this file contains the line*

```
58.3815412948909298 -8.8063413553403507 -5.6290446018702553
```

whereas the corresponding line in the project `cand` *is*

```
58.3815412948909298         2.1858505500938747         0.3972034177439077
```

In the project `CinderellaNative`, *the 14th component in the* SIGSPEC *result file* `000001.result.dat` *in the project directory is related to the 47th component in the file* `000000.result.dat`. *The two frequencies differ by 0.054, and the Rayleigh frequency resolution of the target dataset is* 0.091, *which is sufficient for a correspondence. In the project* `cand`, *the target sig of 2.387 becomes relevant. Eq.* 3 *yields a frequency resolution 0.042 for this component, which is now too small for a coincidence. The corresponding line in the file* `000001.cd.000000.dat` *consistently indicates that no coincident peak is found for this signal component. In this case,* CINDERELLA *uses a default sig threshold for the comparison data, see "Spectral significance threshold", below.*

Spectral significance threshold

If no coincidence in a comparison dataset is detected for a given target frequency, i. e., if CINDERELLA does not find a valid candidate for this target frequency, a default value is used for the sig in the comparison dataset. The user may specify this CINDERELLA threshold by means of the keyword `defsig` in the `.cnd` file. The same specification may be set for the default csig[1] using the keyword `defcsig`. If one of these keywords is not provided, $\frac{\pi}{4}\log \mathrm{e} \approx 0.341$ is used correspondingly by default. According to Reegen (2007), this is the expected value of the sig for white noise. The underlying assumption is that the residuals after prewhitening of all significant signal components in the comparison dataset represent pure noise, i. e. do not contain any further unresolved signal.

[1]abbreviation for cumulative sig

Example. *The sample* cand *contains the same input as* CinderellaNative *(p. 102), and the file* cand.cnd *contains the two lines*

```
defsig 0
defcsig 1
```

The second row in the output file 000001.cd.000000.dat *of the project* CinderellaNative

```
30.7091991449662061 3.8506295783390758 3.6090130812823817
```

using the default sig and csig threshold $\frac{\pi}{4}\log \mathrm{e}$*, because the comparison dataset does not contain a coincident signal component. The second row in the corresponding file of the project* cand *is*

```
30.7091991449662061 4.1917236667995361 2.9501071697428420
```

Since the default sig is lower in the project cand*, the resulting conditional sig is higher. On the other hand, the default csig is higher, which causes the resulting conditional csig to drop down.*

6.2. Amplitude transformation

The assumption that instrumental and environmental artifacts use to be additive in terms of intensity may create needs to transform amplitudes in mag from the comparison into the target spectra, if the conditional CINDERELLA mode is applied. The amplitude transformation is only performed to obtain conditional sigs.

Three different strategies to adjust comparison amplitudes are offered, according to the specifications in the .cnd file.

- In photometry, the photon statistics may be employed to transform comparison into target amplitudes, if the mean magnitudes of the stars are known (Reegen et al. 2008). If the keyword transam:mean is provided in the .cnd file, CINDERELLA uses the mean observables $\langle m_C \rangle$, $\langle m_T \rangle$ of the comparison and target time series, respectively, to transform the comparison amplitude A_C into a target amplitude A_T according to

$$A_T = 2.5 \log \left[1 + \frac{10^{-0.4(\langle m_C \rangle - A_C)} - 10^{-0.4 \langle m_C \rangle}}{10^{-0.4 \langle m_T \rangle}} \right] . \qquad (4)$$

- If the keyword transam:rms is specified in the .cnd file, CINDERELLA interprets the residual rms errors σ_C, σ_T of the comparison and target time series[2], respectively, as measures of the photon noise levels and evaluates the transformed amplitude according to

$$A_T = 2.5 \log \left[1 + \frac{\sigma_C^2}{\sigma_T^2} \left(10^{0.4\,A_C} - 1 \right) \right] . \tag{5}$$

- The keyword `transam:ppsc` in the `.cnd` file causes CINDERELLA to use Eq. 5 employing residual point-to-point scatter instead of residual rms error for both σ_C and σ_T.
- If the keyword `transam:none` is specified in the `.cnd` file, no amplitude transformation is performed at all, i. e., CINDERELLA assumes $A_T = A_C$.

Example. *The sample project* `transam-mean` *contains the same input as the project* `CinderellaNative` *(p. 102). The time series data are considered to represent millimag photometry. The comparison dataset is assumed to refer to a 5 mag star, whereas the target datasets shall correspond to a 15 mag star. The resulting time series input files are* `000000.transam-mean.dat` `000001.transam-mean.dat` *to* `000008.transam-mean.dat`*. The keyword*

```
transam:mean
```

in the file `transam-mean.cnd` *forces* CINDERELLA *to employ the mean magnitudes of the datasets for the amplitude transformation.*

Example. *The sample project* `CinderellaNative` *contains an amplitude transformation based on the rms residual, which is the default method.*

Example. *The sample project* `transam-ppsc` *contains the same input as* `CinderellaNative` *(p. 102). The line*

```
transam:ppsc
```

in the file `transam-ppsc.cnd` *forces* CINDERELLA *to employ the residual point-to-point scatters of the datasets for the amplitude transformation.*

Example. *The sample project* `transam-none` *contains the same input as* `CinderellaNative` *(p. 102). The line*

```
transam:none
```

in the file `transam-none.cnd` *switches off the amplitude transformation.*

[2] with all significant signal prewhitened

7. Composed Mode

The composed sig is a measure of the probability that two coincident signal components occur in two different datasets. This implements a logical 'and', providing an answer to the question, "What is the probability that two different datasets show coincident signal components with given amplitudes and sigs?"

The composed mode is useful for, e. g., photometry of the same star in different filters, or if two short datasets of the same object obtained in different years are examined.

Note that the composed sig in the SIGSPEC result files (see SIGSPEC manual) is consistently defined, but applies to the set of significant signal components displayed in the file, whereas CINDERELLA refers to the composed sig of signal components found in two or more different datasets.

Contrary to the candidate selection procedure in conditional mode (p. 110), the frequency interval between the lowest and the highest frequency found in all target datasets is scanned in steps defined by half the frequency resolution (p. 110). For each of the frequencies under consideration, CINDERELLA computes a composed sig, basically following the introduction by Reegen et al. (2008). Since CINDERELLA's composed mode takes into account all signal components in all datasets, statistical weights have to be introduced that put more emphasis to signal frequencies closer to the frequency under consideration. Hence the composed sig csig (A_n) (annotation by Reegen et al. 2008) assigned to an arbitrary frequency f is evaluated according to

$$\log\left[1 - 10^{\mathrm{csig}(A_n)}\right] = \frac{1}{N}\sum_{n=1}^{N} \mathrm{e}^{-\frac{1}{2}\left[\frac{f-f_n}{\min(\delta f_n)}\right]^2} \log\left[1 - 10^{-\mathrm{sig}\,(A_n)}\right] . \qquad (6)$$

In this context, the total number of signal components in all target datasets is denoted N, f_n referring to the frequency of one of these signal components. The minimum frequency resolution $\min(\delta f_n)$ incorporates the definition by Eq. 3.

An interpolation loop is used to exactly identify the maxima of this composed sig, which are written to the output file.

7.1. Output for composed mode

The calculation of the composed sig is applied to all target datasets at once. A file `cp.dat` is generated, which contains the composed sigs of all target datasets. The three columns refer to:

1. target frequency [inverse time units],
2. composed sig for all target datasets,

3. composed csig for all target datasets.

The composed sigs are also calculated for the conditional sigs, i. e., the files #target#.cd.dat (see "Multi-comparison output", p. 109), and written to a file cc.dat. The column format is the same as for the file cp.dat.

Example. *For the sample project* CinderellaNative, *the project directory contains a file* cp.dat *with the composed sigs (and csigs) for all target datasets (provided by the* SIGSPEC *result files* 000001.result.dat *to* 000008.result .dat*). Furthermore, a file* cc.dat *is found in the project directory. It contains the composed sigs (and csigs) applied to the conditional ones for all target datasets, i. e., the composed sigs are evaluated using the multi-comparison output files generated by the conditional mode,* 000001.cd.dat *to* 000008.cd .dat.

8. Keywords Reference

This section is a compilation of all keywords accepted by CINDERELLA. A brief description of arguments and default values is given. The type of argument is provided by either <int>, <double>, or <string>, and default values are given in parentheses, e. g. (2). Empty parentheses indicate that there is no default setting.

`col:obs <int> (2)`

observable column index (unique), starting with 1, SIGSPEC manual, p. 13

`col:time <int> (1)`

time column index (unique), starting with 1, SIGSPEC manual, p. 13

`col:weights <int>`

weights column index (also multiple), starting with 1, SIGSPEC manual, p. 13

`comp <int> (all except start index)`

specification of time series input files to be regarded as comparison datasets, p. 107

`defcsig <double>` ($\frac{\pi}{4} \log e \approx 0.341$)

threshold to be used for the csig, if no coincidence is found in a comparison dataset, p. 111

`defsig <double>` ($\frac{\pi}{4} \log e \approx 0.341$)

threshold to be used for the sig, if no coincidence is found in a comparison dataset, p. 111

`deftype <target/comp/skip> (target)`

specifies the type of dataset to be assigned to a time series by default, p. 107

`skip <int> ()`

specification of time series input files not to be taken into consideration, p. 107

`target <int> (start index)`

specification of time series input files to be regarded as target datasets, p. 107

`tol <double> (0)`

CINDERELLA frequency tolerance parameter, p. 110

`transam:mean`

amplitude transformation using the mean observable for photon statistics, p. 112

`transam:none`

no amplitude transformation at all, p. 113

`transam:ppsc`

amplitude transformation using the point-to-point scatter of the residual observable for photon statistics, p. 113

`transam:rms` (default)

amplitude transformation using the rms residual observable for photon statistics, p. 112

9. Online availability

The ANSI-C code is available online at `http://www.sigspec.org`. For further information, please contact P. Reegen, `peter.reegen@univie.ac.at`.*

Acknowledgments. PR received financial support from the Fonds zur Förderung der wissenschaftlichen Forschung (FWF, projects P14546-PHY, P17580-N2) and the BM:BWK (project COROT). Furthermore, it is a pleasure to thank M. Gruberbauer (Univ. of Vienna), D. B. Guenther (St. Mary's Univ., Halifax), M. Hareter, D. Huber, T. Kallinger (Univ. of Vienna), R. Kuschnig (UBC, Vancouver), J. M. Matthews (UBC, Vancouver), A. F. J. Moffat (Univ. de Montreal), D. Punz (Univ. of Vienna), S. M. Rucinski (D. Dunlap Obs., Toronto), D. Sasselov (Harvard-Smithsonian Center, Cambridge, MA), L. Schneider (Univ. of Vienna), G. A. H. Walker (UBC, Vancouver), W. W. Weiss, and K. Zwintz (Univ. of Vienna) for valuable discussion and support with extensive software tests. I acknowledge the anonymous referee for a detailed examination of both this publication and the corresponding software, as well as for the constructive feedback that helped to improve the overall quality a lot. Finally, I address my very special thanks to J. D. Scargle for his valuable support.

References

Kallinger, T., Reegen, P., Weiss, W. W. 2008, A&A, 481, 571

Reegen, P. 2005, in *The A-Star Puzzle*, Proceedings of IAUS 224, eds. J. Zverko, J. Ziznovsky, S.J. Adelman, W.W. Weiss (Cambridge: Cambridge Univ. Press), p. 791

Reegen, P. 2007, A&A, 467, 1353

Reegen, P. 2011, CoAst 163, 3

Reegen, P., Gruberbauer, M., Schneider, L., Weiss, W. W. 2008, A&A, 484, 601

Zwintz, K., Marconi, M., Kallinger, T., Weiss, W. W. 2004, in *The A-Star Puzzle*, Proceedings of IAUS 224, eds. J. Zverko, J. Ziznovsky, S. J. Adelman, W. W. Weiss (Cambridge: Cambridge Univ. Press), p. 353

Zwintz, K., Weiss, W. W. 2006, A&A, 457, 237

*Please contact Michael Gruberbauer, `mgruberbauer@ap.smu.ca`.

Comm. in Asteroseismology
Volume 163, 2011

Combine User's Manual

P. Reegen

Institut für Astronomie, Türkenschanzstrasse 17, 1180 Vienna, Austria

Abstract

COMBINE is an add-on to SIGSPEC and CINDERELLA. A SIGSPEC result file or a file generated by CINDERELLA contains the significant sinusoidal signal components in a time series. In this file, COMBINE checks one frequency after the other for being a linear combination of previously examined frequencies. If this attempt fails, the corresponding frequency is considered "genuine". Only genuine frequencies are used to form linear combinations subsequently. A purely heuristic model is employed to assign a reliability to each linear combination and to justify whether to consider a frequency genuine or a linear combination.

1. What is COMBINE?

COMBINE performs an iterative analysis of the frequencies in a `result.dat` file generated by SIGSPEC (Reegen 2005, 2007, and this issue, p. 3 – 97) or one of the output files generated by CINDERELLA (Reegen et al. 2008; this issue, p. 99 – 117). The input file type is detected automatically.

If the attempt to interpret a given frequency as a linear combination fails, this frequency is considered genuine. Only genuine frequencies are used to form linear combinations in the subsequent iterations. The decision whether to accept a linear combination is drawn using a mathematical model to assign an equivalent spectral significance (hereafter abbreviated by 'sig') to a linear combination. This equivalent sig is compared to the sig of the given signal component, and only if it is high enough, the program adopts it.

If there is more than one linear combination available, COMBINE picks the one with the highest equivalent significance.

The underlying model leading to equivalent sigs and the reliabilities of linear combinations is purely heuristic and attempt to mimic the examination by an experienced person.

2. Input

COMBINE is called by the command line

```
combine <infile>
```

where <infile> is the name (or path, if desired) of a SIGSPEC result file or an output file generated by CINDERELLA.

Caution: COMBINE overwrites existing output files!

Furthermore, the user may pass a set of specifications to COMBINE by means of a file <infile>.ini in the same folder as <infile>. For specifications not given by the user, defaults are used.

The file <infile>.ini has to be terminated by a carriage-return character, otherwise the program hangs!

3. How COMBINE Works

For a peak with given frequency and significance, all possible combinations of previously detected genuine frequencies f_k, $k = 1, ..., K$ are computed. K is the maximum number of frequencies in a linear combination. The resulting frequency for a linear combination is

$$f' := \sum_{k=1}^{K} c_k f_k \qquad (1)$$

and shall be compared to a frequency f in the input file.

3.1. Sig vs. csig

If the keyword csig is provided in the file <infile>.ini, the cumulative sig (Reegen 2007, 2011) is used instead of the sig. This keyword does not require any parameters.

3.2. Frequency resolution

The adjustment of the frequency resolution δf is consistent with Eq. 3 in the CINDERELLA manual (see this issue, p. 99 – 117), where the total time interval width T has to be provided by the user, because the time series is not incorporated by COMBINE. Moreover, the user is more flexible if allowed to specify a different value for T. This interval width is provided by means of the keyword `dt` in the file `<infile>.ini`, followed by a floating-point number. The default setting is that COMBINE determines the closest pair of frequencies and uses its inverse frequency spacing as T.

The second parameter, τ, is specified using the keyword `tol`, again followed by a floating-point number, in full consistency with CINDERELLA. The default value is $\tau = 0$, forcing COMBINE to employ the Rayleigh frequency resolution.

The frequency tolerance permits linear combinations where

$$\alpha := |f - f'| \leq \delta f \tag{2}$$

only. The quantity α is the accuracy of a linear combination and provided in the output.

3.3. Limit of harmonic order

The range of harmonic orders is restricted by the parameter N, which is calculated according to

$$N = \text{ceil}\left(\sqrt{\Omega \frac{\text{sig}_k}{\text{sig}_K}}\right) , \tag{3}$$

where sig_k denotes the sig associated to the frequency f_k and sig_K is the sig associated to the last frequency in the input file, f_K. If the keyword `csig` is set, the csig is consistently taken instead of the sig. The parameter Ω is provided by the keyword `order` in the file `<infile>.ini`, followed by a floating-point number. The default value is 1. Given the limit N, the coefficients of a linear combinations are restricted to indices from $-N$ to N according to

$$c_k = -N, \ldots, N \tag{4}$$

.

3.4. Equivalent sig

Each linear combination is assigned an equivalent sig,

$$\text{sig}_{\text{eq}} := \min\left(|c_k|^{-\delta_k}\,\text{sig}_k\right) - \chi \log K , \tag{5}$$

where δ_k denotes the decay parameter provided by the keyword `decay`, and χ is the combination damping, specified using the keyword `cdamp`. Both keywords are followed by floating-point numbers. The default values for both parameters are 1.

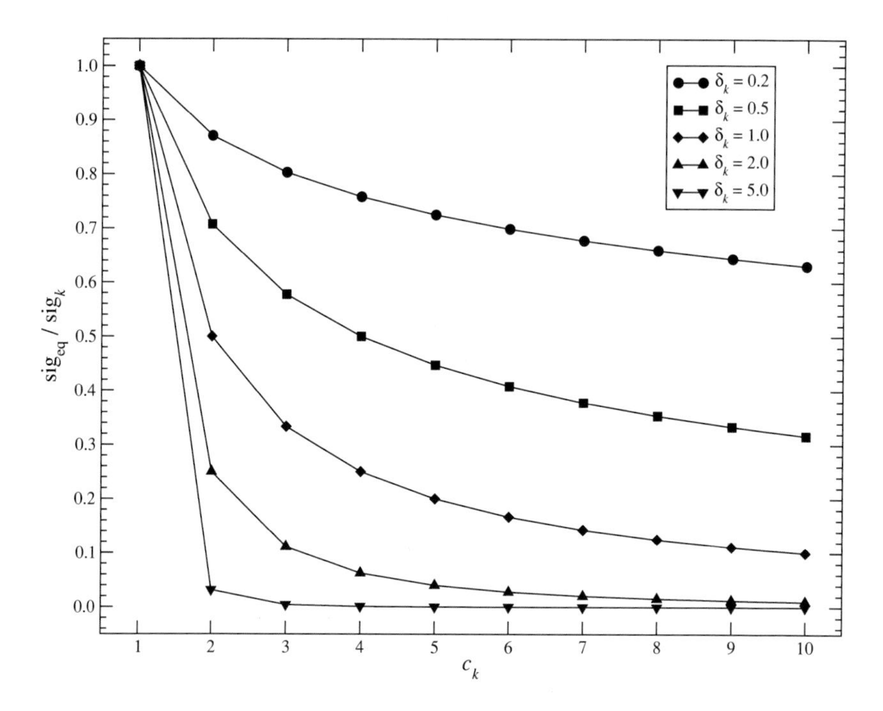

Figure 1: Ratio of equivalent sig over sig of an individual signal component vs. polynomial coefficient c_k associated to the signal component. Five graphs for different values of the decay parameter δ_k are presented.

Fig. 1 displays the relative sig correction with increasing coefficient c_k for five different values of the decay parameter δ_k. Fig. 2 illustrates the correction of equivalent sig with increasing number of components contributing to a linear combination K for five different values of the combination damping χ.

3.5. Reliability and sensitivity

A linear combination is only accepted if the equivalent sig of the combination is high enough compared to the significance of the given peak according to

$$R := \frac{\mathrm{sig}_{\mathrm{eq}}}{\mathrm{sig}_f} \geq S \,, \tag{6}$$

where S is the sensitivity, which can be adjusted by means of the keyword `sens` in the file `<infile>.ini`. The keyword is followed by a floating-point number, and the default value is 0.1. If all examined linear combinations have a reliability below S, the examined signal component is considered genuine.

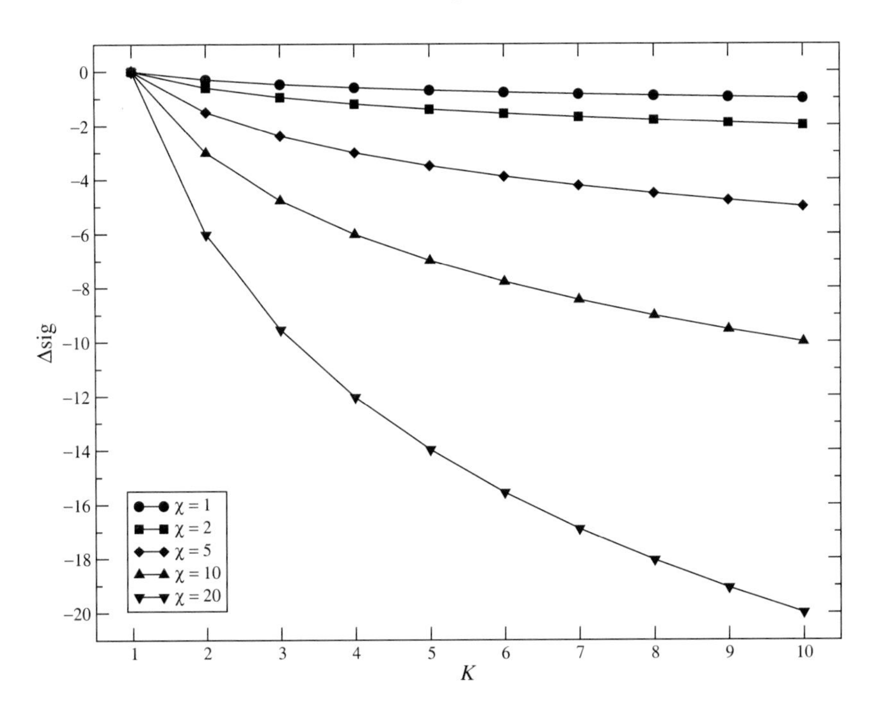

Figure 2: Additive significance correction for a linear combination employing K different signal components. Five graphs for different values of the combination damping χ are presented.

Hence the sensitivity provided by the keyword `sens` permits to directly adjust the number of genuine components in a list of frequencies.

The ratio of sigs, R, is called the reliability of a linear combination and part of the COMBINE output. If multiple combinations are available, the reliability is used to decide which one to pick. This means, COMBINE picks the combination with the highest reliability.

4. Output

Genuine frequencies are assigned identifiers `f#index#`, where `#index#` denotes an integer number starting at 1. According to the number of significant signals present in the file `<infile>`, COMBINE chooses a constant number of digits. For example, if the input file contains from 1 to 9 frequencies, the identifiers for genuine frequencies are `f1`, `f2`, ... If the input file contains from 10 to 99 frequencies, COMBINE enumerates the genuine components `f01`, `f02`, ..., and so on. This format convention applies to the indexing of rows also.

Linear combinations are denoted by the frequency identifiers of the genuine components and appear as a formula: if the frequency under consideration is, e. g., $f_1 + 3f_3 - 2f_{10} - f_{14} - 0.00214$, COMBINE displays it as `f01+3f02-2f10-f14-0.00214` both on the screen and in the output file. In this context, -0.00214 is the frequency accuracy.

The screen output consists of a single line for each signal (i. e., for each row in the input file). COMBINE displays

1. the row index,
2. the linear combination including the frequency accuracy, and
3. the reliability R (Eq. 6).

For genuine frequencies, COMBINE displays only the row index and the frequency identifier. At runtime, the most reliable linear combination identified so far is displayed. If COMBINE finds a "better" solution, the line on the screen is updated.

By default, COMBINE generates an output file `<infile>.cmb`. It contains a row index in the first column, then all information of the input file in the further columns, plus three additional columns at the end:

1. reliability R (Eq. 6)[1],
2. total number of linear combinations within the frequency resolution,
3. the linear combination itself, plus the frequency accuracy. If a frequency is considered genuine, only the frequency identifier is displayed.

For convenience, a second output file `<infile>.gen` is produced by COMBINE. It is truncated to the genuine frequencies only and contains the row index in the first column, then all the information provided in the input file, plus the frequency identifier in the last column. The columns for the reliability and the number of linear combinations within the frequency resolution are omitted. This file provides the opportunity to have all the genuine frequencies available at a glance.

Example.[2] *The sample project* `CombineNative` *contains a list of significant frequencies found in the MOST*[3] *(Microvariability & Oscillations of STars) pho-*

[1] Zero values indicate genuine frequencies

[2] The computation of the sample project `CombineNative` takes 40 minutes on an Intel Core2 CPU T5500 (1.66GHz) under Linux 2.6.18.8-0.9-default i686.

[3] MOST is a Canadian Space Agency mission, jointly operated by Dynacon Inc., the University of Toronto Institute of Aerospace Studies, the University of British Columbia, and with the assistance of the University of Vienna, Austria.

tometry of ζ Oph (Walker et al. 2003, 2004, 2005). According to the input file `result.dat`, *altogether 294 formally significant signal components (sig $>$ 5) were identified.*

The file `result.dat.ini` *contains five keywords:*

```
order 0.2
dt 26
decay 1.5
cdamp 10
sens 0.2
```

The dataset is 26 days long, and the frequencies are provided in cycles per day. Thus COMBINE *will assume a Rayleigh frequency resolution of 0.03846 cycles per day. There is no specification for the frequency tolerance parameter (keyword* `tol`*). Thus the default setting 0 is used.*

Running COMBINE *by typing the command line* `Combine result.dat` *yields a welcome message on the screen.*

```
 CCCCCC                   bb      ii
CC    CC                  bb
CC        ooooo  m mm mm  bb bbb  ii n nnnn   eeeee
CC       oo    oo mm mm mm bbb   bb ii nn    nn ee    ee
CC       oo    oo mm mm mm bb    bb ii nn    nn ee    ee
CC       oo    oo mm mm mm bb    bb ii nn    nn eeeeeee
CC       oo    oo mm mm mm bb    bb ii nn    nn ee
CC    CC oo    oo mm mm mm bb    bb ii nn    nn ee    ee
 CCCCCC   ooooo  mm mm mm b bbbb  ii nn    nn  eeeee


Version 1.0
************************************************************
by Piet Reegen
Institute of Astronomy
University of Vienna
Tuerkenschanzstrasse 17
1180 Vienna, Austria
Release date: August 18, 2009
```

The program finds out that the input file is a seven-column SIGSPEC *result file, determines the number of rows and reads the input data. Note that 295 rows correspond to 294 significant signal components, because the last row in the* SIGSPEC *result file contains information on the residuals (see* SIGSPEC *manual, p. 28).*

```
*** start **************************************************

File result.dat: SigSpec format
rows                                    295
read input file
```

Then the search for linear combinations starts. For each row in the input file, COMBINE *displays the most reliable combination detected so far.*

The first four signal components are found to be genuine. Since the number of signal components is 294, COMBINE *uses a three-digit format for the row indices and frequency identifiers.*

```
row 001: f001
row 002: f002
row 003: f003
row 004: f004
```

For rows 5 and 6 in the input data, the screen output contains the most reliable linear combination (including the frequency accuracy) and the reliability.

```
row 005: 3f001-f002-2f003-f004+0.0284306 0.236585
row 006: 3f001+2f002-f004+0.0136421 0.35803
```

An examination of the output file `result.dat.cmb` *shows that rows* 005 *and* 006 *end with*

```
0.2365853347754522  1 3f001-f002-2f003-f004+0.0284306168856169
0.3580304203945811  2 3f001+2f002-f004+0.0136420746028509
```

These entries refer to the columns added by COMBINE. *The first value is the reliability, the second one is the number of examined linear combinations, and the last column represents the linear combination itself. For row* 005*, there is only one linear combination available within the frequency resolution, for row* 006 *the number of linear combinations taken into account is 2.*

Subsequently, the screen output indicates a fifth genuine frequency.

```
row 007: f005
```

The frequency in row number 8 is 0.02783 cycles per day, which is below the frequency resolution. Thus the component is considered to refer to zero frequency, and in this case, no reliability is evaluated.

```
row 008: 0+0.0278395
```

In the further rows of the input files, no more genuine frequencies are detected.

```
row 009: -f002+f005-0.025485 0.759005
row 010: f001-f002-f004+f005+0.0313392 0.490535
row 011: -f001+f004-0.00275538 1.26888
row 012: f001-f002-f004+f005-0.0295542 0.680494
row 013: -2f001+2f003+f004-0.00567519 0.523911
row 014: -f001+f005+0.024731 1.72772
row 015: 2f002+0.0249392 1.47442
row 016: 2f001-f004-0.0100088 1.70761
row 017: -f001+2f002-0.00217389 1.55951
row 018: f001-f002+0.00824894 3.95466
row 019: f002+f005-0.00668728 1.64167
row 020: 2f002+f003-f005-0.00199182 0.779607
```

It is a remarkable matter of fact that COMBINE *is able to compose all 294 frequencies contained by the input file as linear combinations of no more than five genuine frequencies. However, a different parameter constellation in the configuration file* `result.dat.ini` *can produce completely different output. Note that the time consumption by* COMBINE *dramatically increases with the number of genuine frequencies identified. This is because more genuine frequencies increase the number of possible linear combinations over-proportionally. A list of genuine frequencies only is found in the output file* `result.dat.gen`.

```
5 genuine frequencies found.


Finished.

************************************************************

Thank you for using Combine!
Questions or comments?
Please contact Piet Reegen (reegen@astro.univie.ac.at)
Bye!
```

5. Order of Input Rows

Since COMBINE processes the input file row by row, the order of rows plays a crucial part in the way the analysis is performed. Changing the order of rows in the input file influences the base upon which the linear combinations are formed. Thus, if there are frequencies previously known to be genuine, it is advisable to ensure that they are on top of the input file, if all further frequencies are supposed to be checked for linear combinations of preferrably these components.

Example.[4] *The input of the sample project* `order` *is essentially the same as for* `CombineNative`. *Only the order of rows is slightly modified: the 6th signal component of the file* `result.dat` *in the project* `CombineNative`, *which refers to the orbit frequency of the MOST spacecraft, appears now on top. This re-ordering forces* COMBINE *to consider* $14.188\,\mathrm{d}^{-1}$ *genuine. Also the configuration file* `result.dat.ini` *is the same as for the project* `CombineNative`.

Again, there is a base of five genuine frequencies three of which are identical to the project `CombineNative`, *namely 5.182, 2.675 and 3.055 cycles per day. The two genuine signal components at 6.722 and 7.193 cycles per day are replaced by 14.188 and 0.0697 cycles per day.*

[4] The computation of the sample project order takes 40 minutes on an Intel Core2 CPU T5500 (1.66GHz) under Linux 2.6.18.8-0.9-default i686.

6. Rejecting Unwanted Linear Combinations

Moreover, the user may indicate unwanted signal components in the input file `<infile>` by applying a minus sign to the corresponding frequencies. COMBINE reacts with a corresponding change of the sign for the reliability. If the user additionally provides the keyword `reject` in the file `<infile>.ini`, all rows are rejected from the output file `<infile>.cmb` for which the most reliable linear combination contains one or more unwanted frequencies.

The screen output contains linear combinations incorporating unwanted frequencies at runtime. To indicate such unwanted combinations, the reliability is displayed as a negative value. If the examination of an input line finishes with the "best" linear combination containing an unwanted frequency, the corresponding line is removed from the screen output.

Example.[5] *The input of the sample project* `reject` *is the same as for* `order`, *with a minus sign for the first frequency of 14.188 cycles per day, which represents the orbit of the MOST spacecraft. The file* `result.dat.ini` *contains an additional line,*

```
reject
```

The combination of this keyword and the negative sign for the first signal component in the input file forces COMBINE *to reject all linear combinations incorporating the frequency 14.188 cycles per day from the output file* `result.dat.cmb`. *In the screen output, such linear combinations are indicated by a negative reliability, e.g.*

```
row 005: f001+3f002+2f003+0.0136421 -0.325575
```

This entry is visible at runtime, but vanishes from the screen output when the calculations for row 006 start.

7. Keywords Reference

This section is a compilation of all keywords accepted by COMBINE. A brief description of arguments and default values is given. If an argument is required, it is indicated by `<double>`, and default values are given in parentheses, e.g. `(1)`.

```
cdamp <double> (1)
```

combination damping, e.g. reduction of reliability of a linear combination with increasing number of components employed, p. 121

[5]The computation of the sample project `reject` takes 40 minutes on an Intel Core2 CPU T5500 (1.66GHz) under Linux 2.6.18.8-0.9-default i686.

`csig`

forces COMBINE to use csig instead of sig, p. 120

`decay <double> (1)`

decay of reliability assigned to a frequency multiple for increasing harmonic order, p. 121

`dt <double> (auto)`

total time interval of the time series, defining the Rayleigh frequency resolution. By default, COMBINE determines the Rayleigh frequency resolution as the frequency spacing of the closest pair of frequencies found in the input data, p. 121.

`order <double> (auto)`

parameter restricting the range of harmonics of individual frequency components to be employed to form linear combinations, p. 121

`reject`

activates the rejection of unwanted linear combinations. The user may indicate unwanted frequencies by a minus sign in the input file `<infile>`. If this keyword is set, COMBINE automatically suppresses the output of those signal components for which the most reliable linear combination incorporates such an unwanted frequency, p. 128.

`sens (0.1)`

reliability limit to be exceeded in order to accept a linear combination, adjusts the number of genuine components in a frequency list, p. 122

`tol <double> (0)`

COMBINE frequency tolerance parameter, p. 121

8. Online availability

The ANSI-C code is available online at `http://www.sigspec.org`. For further information, please contact P. Reegen, `peter.reegen@univie.ac.at`.*

Acknowledgments. PR received financial support from the Fonds zur Förderung der wissenschaftlichen Forschung (FWF, projects P14546-PHY, P17580-N2) and the BM:BWK (project COROT). Furthermore, it is a pleasure to thank M. Gruberbauer, M. Hareter, D. Huber, D. Punz (Univ. Vienna), G. A. H. Walker (UBC, Vancouver), and W. W. Weiss (Univ. Vienna) for their help. I acknowledge the anonymous referee for a detailed examination of both this publication and the corresponding software, as well as for the constructive feedback that helped to improve the overall quality a lot. Finally, I address my very special thanks to J. D. Scargle for his valuable support.

References

Kallinger, T., Reegen, P., Weiss, W. W. 2008, A&A, 481, 571
Reegen, P. 2005, in *The A-Star Puzzle*, Proceedings of IAUS 224, eds. J. Zverko, J. Ziznovsky, S.J. Adelman, W.W. Weiss (Cambridge: Cambridge Univ. Press), p. 791
Reegen, P. 2007, A&A, 467, 1353
Reegen, P. 2011, CoAst 163, 99
Reegen, P., Gruberbauer, M., Schneider, L., Weiss, W. W. 2008, A&A, 484, 601
Walker G., Matthews J., Kuschnig R., et al. 2003, PASP, 115, 1023
Walker G. A. H., Matthews, J. M., Kuschnig, R., et al. 2004, BAAS, 36, 1361
Walker G. A. H., Kuschnig, R., Matthews, J. M., et al. 2005, ApJ, 623, L145

*Please contact Michael Gruberbauer, `mgruberbauer@ap.smu.ca`.